Laurie,   2/22/2019

This would have never happened without your assistance; and that first phone call.

The Goodman family is forever in my heart as it has been all of my life.

Thank you for a lifetime of generosity, friendship, love + support.

— And thanks for opening all those gates! —

Cowboy Up!

Robert

Robert E. Forbis Jr.

Altered Policy Landscapes

Fracking, Grazing, and the Bureau of Land
Management

 Springer

Robert E. Forbis Jr.
Department of Political Science
Texas Tech University
Lubbock, TX, USA

ISBN 978-3-030-04773-3   ISBN 978-3-030-04774-0   (eBook)
https://doi.org/10.1007/978-3-030-04774-0

Library of Congress Control Number: 2018966337

This Springer imprint is published by the registered company Springer Nature Switzerland AG
The registered company address is: Gewerbestrasse 11, 6330 Cham, Switzerland

Foreword

Robert and I first met when we were doctoral candidates, he at the University of Utah and I at Colorado State University. Sharing research interests, nontraditional career paths, value systems, and theoretical and methodological perspectives, we invariably presented our research on the same Western Political Science Association conference panels focusing on western US environmental and energy policy. The inattention to energy policy, especially western oil and gas issues, was vividly clear as panelists frequently outnumbered those in the audience. Ironically, that setting served as an incubator to test ideas free from pressure and unnecessarily scathing, ego-busting academic critique. As the salience and political conflict over energy development in the Rocky Mountain West grew, so did the audiences and our respective research agendas. It was exciting! We were studying a western US energy boom while it was happening and prior to the unconventional oil and gas energy boom that swept other regions beginning in the late 2000s.

Importantly, this book not only vividly chronicles that boom, the key actors, and political outcomes but also substantively adds to our understanding of environmental federalism and energy policy through the lens of subgovernment theory. Many political scientists and policy scholars assert that natural resource conflicts, in particular, and policy-making, in general, are better characterized by open systems, issue networks, and pluralistic bargaining. This book effectively argues and demonstrates that policy subgovernments or iron triangles still exist and the theory remains viable, testable, and explanatory. How can former allies like ranching and energy, who both have long dominated and enjoyed the benefits of generous federal public land and mineral estate management policies, turn on each other? What conditions are necessary to alter domination of a policy subgovernment; create unusual political alliances between ranchers, property rights advocates, and environmental groups; fundamentally alter vast expanses of western landscapes; and create both federal and state shifts in energy policy? This book answers those questions using comparative case study methods and a fantastically thick, rich qualitative analysis within the umbrella of subgovernment theory.

Arguably, the fight over how to manage public lands in the American West has always been central to state and local politics, but public lands' battles also serve as an exemplar of the difficulties in the implementation of environmental federalism and bureaucratic control. Too frequently, the historical context and statutory framework underpinning political conflicts and bureaucratic control are ignored or given scant attention, but not here. Chapter 2 covering the legal history and key statutes that enabled western colonization while separating surface from subsurface mineral ownership rights is a must-read. How bureaucratic implementation of the split estate impacts present public and private land uses is paramount to understanding this modern energy policy change. Strikingly, the Department of Interior's Bureau of Land Management (BLM) follows its complex multiple-use statutory mandates within the more immediate context and directives of executive branch policy preferences. Thus, this study provides an insightful contribution to the literature on presidential power and how effectively their tools can be used to fundamentally alter surface and subsurface land use management policy through the bureaucracy (Chap. 3). Notably, congressional committee and powerful interest group support compliment and further empower executive control over the bureaucracy, fundamental precepts of subgovernment theory.

We learn, remember, and are enthralled by stories. I am simultaneously fascinated yet exasperated by the story of Manifest Destiny, colonization, resource exploitation, land use evolution, and the scars and voids these changes, driven by policy, have left behind. The true strength of this book resides within the stories told by those who have been interviewed. They tell the modern story and recollection of events, while the statutes and archival documents tell the historical story. Together they weave a powerful, vivid, and passionate narrative from unique perspectives. BLM appointees, a Colorado state representative, Wyoming Speaker of the House, presidents of energy industry associations, cattleman's association lobbyists, and ranching association presidents tell these stories with alacrity, incredulity, frustration, fervor, insight, and candor. The simple fact that both ranching and energy industry representatives would agree to be interviewed on the record is a testament to the networking, persistence, and communication abilities of Dr. Forbis. As the book explores in great detail, the hegemonic narrative of ranching on public lands remains both mythically and politically powerful, but this narrative has been undermined and supplanted by a narrative rooted in domestic energy independence, technological advances, energy security, and economic opulence.

Turning historic allies, like ranching and energy, into foes is in itself an interesting story. Turning a bureaucracy from privileging one industry over another provides another layer of complexity. Turning from a federal bureaucracy not given the tools and broad statutory mandates to solve complex public and private land use disputes to the states for problem resolution is a necessary and logical outcome and opportunity afforded by federalism. Turning the pages of this book illuminates the genesis, complexity, and evolution of this ongoing, intractable, and national energy policy debate.

Associate Professor, Dual Appointment,   Andrew R. Kear, Ph.D.
Department of Political Science,
Department of Environment and Sustainability,
School of Earth, Environment and Society
Bowling Green State University,
Bowling Green, OH, USA

Acknowledgments

This book has benefited from the insights and criticisms of many people. Early elements of the critique presented in this manuscript were submitted in a variety of forms to a variety of professors, colleagues, and friends. All members and faculty of the Department of Political Science, Public Policy and Administration program at the University of Utah have generously supported me throughout this research project, and I am grateful for their continued support as I wander down the perilous road of academic exploration.

I am particularly grateful for the guiding insight of my dissertation committee chair, Dr. Daniel McCool. I am equally grateful for the collective knowledge of my dissertation committee members, Dr. Richard Green, Dr. Peregrine Schwartz-Shea, Dr. Daniel Levin, and Dr. Sandi Parkes. Each of these remarkable professors has left the distinct imprint as mentors and friends not only on this work but on me as well.

I am very grateful to all the people who agreed to participate in this research project. Thanks go to Pat Shea, former Director of the BLM under President William J. Clinton; Don Simpson, State Director of Wyoming BLM; Larry Claypool, Deputy State Director of Minerals and Lands Wyoming BLM; Lynn Rust, Deputy State Director of Minerals and Lands Colorado BLM; and Tony Herrell, Deputy State Director of Minerals and Lands New Mexico BLM; Rebecca Watson, former Assistant Interior Secretary for Lands and Mineral Management under former President George W. Bush; and an Unnamed DOI political appointee under former President George W. Bush.; Colorado State Representative Ellen Roberts (R-Dist.59); Kathleen Sgamma, Director of Government Affairs for the Independent Petroleum Association of Mountain States; Bob Gallagher, former President of the New Mexico Oil and Gas Association; Stan Dempsey, President of the Colorado Petroleum Association; Bruce Hinchey, President of the Petroleum Association of Wyoming (PAW) and former Speaker of the House, State of Wyoming Legislature; Laurie Goodman and John Vincent of the Landowners Association of Wyoming; Jim Magagna of the Wyoming Stock Growers Association; and Caren Cowan of the New Mexico Cattle Growers Association. Thanks go to all the other folks, who at one time or another shared their thoughts with me as I sought greater understanding and meaning for the research project.

Sincere appreciation and a special note of thanks to MPA graduate student and APA expert Lindsay Heightman of Idaho State University for taking the time and having the patience to review, edit, and account for all my resources. Thanks to Dr. Donna Lybecker of Idaho State University for early review and editing of my work. Special appreciation goes to my new colleagues at Texas Tech University Department of Political Science for their patience and amazing support as I chased down an ever-moving/ever-changing press publication promise and now finally, a firm contract and with it, a deadline.

I would be remiss were not to thank Dr. Andrew Kear of Bowling Green State University for lending a critical voice to the revision process as well as an incredible foreword to the book. As Dr. Kear and I were working through our respective doctoral studies, he at Colorado State University and I at the University of Utah, we discovered that we were exploring a fascinating, emerging, political conflict that was unfolding right before our very inquisitive eyes. We were the first. Often unrecognized, ignored, and dismissed, we knew that the expansion of hydraulic fracturing across the West was a significant phenomenon that deserved academic attention. We weren't wrong. Though, we both understand that there's nothing more humbling than speaking to the exotic nature of cows on a conference panel intended to discuss environmental tourism. I am forever indebted to Andy's friendship and wise counsel.

Thanks to Springer Press for offering me the opportunity to bring this story to light. Their willingness to assist me in the process of revision, editing, and helpful suggestions has made this work even more impactful. Specific thanks to Zachary Romano for reading the manuscript and understanding its importance. Thanks as well to Krishnan Sathyamurthy for unending patience as the revised manuscript was being submitted for his careful review.

Thanks go to my parents, Bob and Pat, for instilling in me the work ethic to see this project through to its conclusion. Thank you to my brothers, Tim, Rick, and Mike. Thanks go to members of my amazing PhD cohort, Dr. Steve Nelson, Dr. Jennifer Robinson, and Utah State Representative Dr. Jennifer Seelig, for their continued support and friendship. Thanks to all my wonderful friends for patiently listening to my endless rants as I attempted to clearly express my thoughts through the research, writing, and revision process. Special thanks go to my sons, Sean and Ian Forbis, for inspiring me to earn my education.

Thanks and prayers go to my late grandmother, Pauline Buck; her love and the values she instilled in me continue to motivate me to this day. Her belief in me never wavered. Her unwavering confidence has sustained me throughout my life. Finally, I must thank Lisa Natale, my wife, my friend, and my partner in all things academic and otherwise, for her love, patience, and understanding. I do not believe I could have made my life's ambition a reality without her confidence, support, and sacrifice. Thank you Lisa, and yes, I owe you.

I, of course, am solely responsible for any errors or shortcomings that may remain in this research.

Contents

Chapter 1
Introduction

Abstract In 2000 as the BLM responded to the external pressures of presidential pressure, and often political appointee actions emphasizing domestic energy production, a conflict emerged between traditional subgovernment interest alliance of ranching and energy development. As an unintended consequence of the expansion of domestic energy production, energy development expanded onto split-estates— where property rights are severed between a privately owned surface estate and the federally owned and managed subsurface mineral estate—triggering conflicts that resulted in the disruption of the traditional subgovernment alliance between ranching and energy interests. The conflict grew over time as the energy industry—as well as the George W. Bush administration—sought to displace the ranching industry's historical domination of the BLM and its land-use policy subgovernment.

Keywords Bureau of Land Management · Subgovernment · Ranching · Energy · Public policy · Executive power · Political conflict

Literature analyzing the Bureau of Land Management (BLM) land-use policymaking environment is varied. The vast majority of that literature emphasizes that the BLM is an agency fraught with conflict and competition among stakeholders seeking to control the policy agenda (Cawley, 1993; Clarke & McCool, 1996; Culhane, 1981; Davis, 1997a; Donahue, 1999; Foss, 1960; Klyza, 1996; Knight, Gilgert, & Marston, 2002; Merrill, 2002; Nie, 2008; Smith & Freemuth, 2007; Starrs, 1998; Wilkinson, 1992). This book adds to the existing literature by arguing that the BLM has shifted from a rancher-dominated agency to an energy-dominated agency. The data collected for this research effort also significantly advances the analytical power of subgovernment theory and illustrates the methodological utility of process tracing.

In doing so the book addresses the historical shift by advancing three primary arguments: (1) changes in the executive branch led to changes in domestic energy policy; (2) changes in domestic energy policy triggered heightened conflict and competition between formerly allied, strong, and resource-rich members in a public lands' subgovernment; and (3) heightened conflict and competition between former subgovernment allies led to a shift in policy control of a public lands' subgovernment.

© Springer Nature Switzerland AG 2019
R. E. Forbis Jr., *Altered Policy Landscapes*,
https://doi.org/10.1007/978-3-030-04774-0_1

The arguments presented here are also used as a means of improving political science's understanding of subgovernment theory by advancing its analytical power in two ways: (1) clarifying the impact executive decision-making has on subgovernments and establishing its potential for detrimental effects on established subgovernment alliances and (2) identifying the conditions under which a strategically competitive behavior between two competing subgovernment actors seeking political and policymaking advantage occurs. The previously mentioned arguments and the use of subgovernment theory specifically seek to *identify conditions* under which subgovernment actors *strategically respond* to a *political conflict*. Finally, the method of data collection and analysis in support of the findings illustrates the methodological utility of process tracing by (1) conducting a comprehensive review of archival documentation as a means of indicating the causal relationship between executive action, the mechanism for disruption and conflict, and the resulting shift in subgovernment dominance and (2) conducting elite interviews as a means to trace, contextualize, and confirm the causal processes, mechanism, and shift in subgovernment dominance indicated by the analysis of the archival documentation.

The Conflict

In 2000 as the BLM responded to the external pressures of presidential pressure, and often political appointee actions[1] emphasizing domestic energy production, a conflict emerged between traditional subgovernment interest alliance of ranching and energy development. As an unintended consequence of the expansion of domestic energy production, energy development expanded onto split-estates—where property rights are severed between the privately owned surface estate and the federally owned and managed subsurface mineral estate[2]—triggering conflicts that resulted in the disruption of the traditional subgovernment alliance between ranching and energy interests (Hardin & Jehl, 2002; Miller, Hamburger, & Cart, 2004; Wilkinson, 2005). This conflict grew over time as the energy industry—as well as the George W. Bush administration—sought to displace the ranching industry's historical domination of the BLM and its land-use policy subgovernment.

Evidence of energy's emerging dominance of the BLM and ranching's loss of influence over domestic energy development since 2000 expanded beyond the traditional confines of public lands and increasingly encroached upon the surfaces

---

[1] Note: See The Wilderness Society, n.d.; see also the Bureau of Land Management [BLM], n.d. for information regarding Congress's directive to BLM, as well as other information concerning federal action and split-estates, best management practices, rights, and responsibilities. See also Energy Policy Act of 2005, Split-Estate Federal Oil and Gas Leasing and Development Practices, § 1835, 119 Stat. 594.

[2] Note: See generally BLM, n.d.; the Bureau of Land Management estimates 58 million Western split-estate acreage (nonfederal surface/federal minerals) and seven million acres of non-Western split-estate acreage. See also generally Environmental Working Group, 2004.

of privately owned ranchlands.[3] If the ranching industry still dominated the BLM and its land-use policies, those same interests could have relied on their control of the policy subgovernment to protect their interests. Ample evidence suggests that this did not happen.

Instead, as the energy industry's emerging domination of the BLM's land-use policymaking became evident to ranchers, they began to form new organizations and alliances and sought the protection of their respective state legislatures.[4] These actions led to fiercely competitive political responses between members of the traditionally allied ranching and energy groups for control of the BLM's subgovernment and its agenda of land-use policymaking (Eilperin, 2006; Lofholm & McGuire, 2006; www.denverpost.com, 2006). The legislative battles suggest that, as the level of conflict and competition intensified, control over the BLM's land-use policymaking shifted from ranching interests to energy development interests.[5]

These conflicts are chronicled and analyzed in three Western states: New Mexico, Colorado, and Wyoming. These conflicts illustrate how the strategic behavior of subgovernment actors competed for dominance of the BLM's policy subgovernment. Analysis of these conflicts strongly suggests that a shift in stakeholder control of the BLM's policy agenda has occurred. Subgovernment theory has been criticized as a simple descriptive device for identifying causal relationships between actors and the strategies they employ to dominate the policy setting (McCool, 1989, 1990, 1995, 1998). This research project advances an analytical version of subgovernment theory by establishing clear causal linkages among elite, politically powerful, decision-making actors in an established policy subgovernment. More specifically, it advances the analytical power of subgovernment theory by exploring how the BLM's land-use subgovernment actors responded to an unintended political conflict resulting from executive branch actions that concluded with a shift in the control over the policymaking setting.

Subgovernment Theory

The so-called iron triangles are the classic model for describing policy subgovernments (Cater, 1964; Freeman, 1965; Lowi, 1979; McConnell, 1966; McCool, 1989, 1990, 1995, 1998). In the iron triangle model, relationships between interest groups, agency bureaus, and congressional subcommittees are described as "mutually supportive and harmonious" (Kelso, 1995). As suggested by the model's name, the iron nature of these mutually supportive and harmonious relationships means they resist

---

[3] Note: See generally BLM, n.d., for a summary discussion of energy expansion to private lands.

[4] Note: See generally Earthworks, n.d., for a summary description concerning previously implemented surface owner protection or damage compensation laws in Western states excluding Utah.

[5] Note: See generally Western Organization of Resource Councils, n.d., "Supporting Declarations," of first person accounts to the "Oil and Gas Industry Responsibility Petition" to the Dept. of the Interior and BLM.

the influence of other actors. Academic research has long used natural resource policy as a lens into subgovernment behavior (Cahn, 1995; Castelnuovo, 1998; Cawley, 1993; Hage, 1994; Merrill, 2002; Yandle, 1995). And as public policy research—particularly environmental policy research—has grown more complex over time and these iron-clad relationships are now described as "open systems" (Kelso, 1995), previously closed policy domains are now described as porous and susceptible to the influence of competing players (Kelso, 1995). The increasingly complex relationships between policy actors operating within such "open systems" have been conceived of and tested by advocates of multiple models including laissez-faire pluralism (Dahl, 1967; Truman, 1971), elite pluralism (Lowi, 1979), issue networks (Heclo, 1978), advocacy coalitions (Sabatier & Jenkins-Smith, 1999), policy streams (Kingdon, 1984), and punctuated equilibrium (Baumgartner & Jones, 1993).

The theoretical shift from the simplicity of iron triangles to the complexity of open systems is illustrated in the historical domination of natural resource policy-making by large user interest groups. Charles Wilkinson (1992) describes the dominance of singular, large user interest groups in a variety of natural resource policy settings. Critically assessing nineteenth- and early twentieth-century natural resource laws, policies, and ideas, Wilkinson argues that "natural resources are governed by what I have come to think of as the 'lords of yesterday'" (Wilkinson, 1992, p. xiii). Wilkinson notes that these laws, policies, and ideas were not always irrational but "arose for good reason" at the time of their passage (Wilkinson, 1992, p. xiii).

In each of these natural resource settings, Wilkinson (1992) accounts for a "compounding problem," "the capture of large interests of the laws and policies that comprise the lords of yesterday," and their ability to thwart reform through their substantial political and financial muscle (p. 22). Thus our collective understanding of subgovernment systems and how they operate within various policy settings benefits from researching the historical development of government responses to the use of natural resources. As Charles Davis (1997a) notes:

> …natural resource issues, including water (Ingram, 1990; McCool, 1987), energy development (Jones & Strahan, 1985; Rosenbaum, 1993), agriculture commodities (Browne, 1988), timber harvesting (Clary, 1986), and hardrock mining (Heclo, 1978) …were developed within a distributive policy context…[that] also spawned a protective subgovernment that restricts participation in policy decisions to public agency administrators, legislators, and interest group representatives with shared programmatic concerns. (pp. 7, 87)

Davis then asks, "How can we account for the continuing political strength of the range policy subgovernment in the face of opposition from both environmental groups and advocates of greater efficiency in government" (C. Davis, 1997a, p. 87)?

According to Davis and others, the Taylor Grazing Act of 1934 made grazing the "dominant use" on BLM lands protecting ranching interests. It is, Davis notes, not until the passage of the Federal Land Policy and Management Act (FLPMA) of 1976 amending the Taylor Grazing Act by "replacing the provision identifying livestock grazing as the predominant use of public rangelands" with "the multiple-use management scheme" that ranching interests were confronted with competing land-use interests (C. Davis, 1997a, p. 95). Like Wilkinson, Davis places the Taylor Grazing Act of 1934 in the center of his critique of how ranching's domination of the policy arena endures in the face of FLPMA's significant land-use reforms.

Davis argues that, since the adoption of FLPMA, pro-grazing interests have maintained their policy dominance by maintaining their core support for grazing while gradually expanding to represent the interests of other large resource use interests including, among others, energy companies, and supporting the states' rights and property rights political movements. While Davis notes the role of environmental organizations in advancing changes in rangeland-use policy since the passage of FLPMA, he remains relatively silent on how these groups, as well as the other large resource use interests, accommodate ranching's continued effectiveness in protection grazing as the dominant use of public range lands (D. Davis, 1997b).

Literature focusing specifically on energy policy fares no better explaining the cozy relationship between ranching and energy development interests. Academic inquiries into energy development policy, like those into other public resource policy areas, have been primarily concerned with the conflicting policy environments of energy development interests and environmental protection interests. Rosenbaum (1993) argues that "there is political symmetry to energy and environmental issues. Energy policy is environmental policy by another name" (p. 188). David H. Davis argues that "Four factors may explain the evolution of energy policy on federal lands: (1) interest groups, (2) political partisanship, (3) bureaucratic routines, and (4) economics" (D. Davis, 1997b, pp. 122–124). Both Rosenbaum (1993) and D. Davis's (1997b) arguments are grounded in the symmetrical relationship between energy and environmental issues. Unlike Wilkinson or C. Davis's inquiries, neither Rosenbaum nor D. Davis's research explains the effects of long-dormant legislation or the mutually supportive relationship between ranching and energy development interests on the land-use policymaking subgovernment.

Rosenbaum (1993) concludes that "the risk remains great, as it always has, that national political majorities and interest coalitions will dominate national energy policies and override Western regional interests in the name of the greater national good" (p. 196). Here, he suggests the potential for energy development interests to overwhelm the historic domination of grazing or possibly the doctrine of multiple use. D. Davis (1997b) makes no similar predictions of change but, instead, concludes that a relative stability, even predictability, in the dynamics of energy policy has developed over time (pp. 146–148). Both approaches are overly simplistic because they do not fully account for variance in political conditions and relationship dynamics affecting the subgovernment system of energy resource development.

Other authors describe how the cozy relationships within land-use policy subgovernments have been formalized and institutionalized within government agencies. This body of literature holds that, in order to fully understand public land politics and policy, one must account for different resource management patterns, across administrative settings, within the various resource use policy domains. Here, the key to better understanding the dynamics of resource policy subgovernments benefits from comparing patterns of realignment across governing administrative agencies and their respective resource use subsystems.

Klyza (1996) argues that previous studies have not fully answered the puzzle of "different policy patterns in the same policy area" (p. 6). Klyza (1996) notes that this problem has been inadequately addressed by the previous literature for four primary reasons:

> First, some studies have been descriptive without being theoretical (p. 6, commenting on Wilkinson, 1992) ...some studies have made insufficient comparisons across policy regimes, (p. 6, commenting on Clary, 1986; Durant, 1992) ...[some] studies have focused on only specific agencies rather than the entire policy process (p. 6, commenting on Culhane, 1981; Clarke & McCool, 1996), ...[and] many such studies lack a systematic historical perspective (p. 6, commenting on Clary, 1986; Durant, 1992).

Klyza (1996) suggests that "the key to understanding the puzzle of different policy patterns in public-lands politics is in understanding the foundation of a policy regime [subsystem] and the subsequent politics that emerge from this" (p. 7). Klyza (1996) argues that new policy regimes embody privileged ideologies that guide and constrain the actors within the policy regime. This ideology becomes institutionalized and becomes "very difficult to dislodge, despite challenges from interest groups and agencies supporting other ideas" (1996, p. 7). Yet, no one, including Klyza, has provided evidence to illustrate this point.

Klyza (1996) also argues that "[t]his institutionalization is not forever" and that given the right circumstances, "nonprivileged ideas can be victorious" in dislodging the privileged idea that guides and constrains actors within the policy regime (p. 7). Klyza notes that embedded ideas are extraordinarily difficult to dislodge, even when they are a source of friction. Klyza, however, only suggests possible hypothetical scenarios for triggering disruption. His likely scenarios range from social movements to the rise of a new land management professionalism, agency reorganization, or expansion in administrative agency responsibilities like domestic energy development.

Klyza's (1996) analysis of three cross-cutting privileged ideas and the responses of agencies and interest groups is that "Despite [the] challenges and cracks, the embedded idea has proved difficult to dislodge. In each of the policy regimes, the privileged idea, though tarnished, is still in place" (pp. 141–160). Change, Klyza (1996) concludes, will occur only when a "fundamental change in state and society" transforms prevailing views on the role of government in managing natural resources (p. 159). As noted in the previously reviewed literature, Klyza's conclusions remain primarily focused on the privileged oppositional interplay between resource development interests and environmental interests.

Klyza's (1996) provocative insights suggest a series of important, unanswered questions. Are there cases that illustrate how, when the "right circumstances" exist, "nonprivileged ideas can be victorious?" If such cases exist, does dislodging a privileged idea require a fundamental change in state and society? Might something as simple and direct as a change in the presidency, an executive order, or an executive appointment, or some combination thereof dislodge a privileged idea? Are there cases that demonstrate how events can provoke "oppositional interplay" between interest groups within the land-use subgovernment? And, if such cases exist, might examining how political conflict arising within that subgovernment, between formerly allied interest groups, advance our knowledge of the behavioral dynamics of subgovernments? Could it be that focusing our collective attention on the privileged conflict of development interests and environmental interests has limited our ability to advance subgovernment theory? The research presented in this book seeks to answer these questions.

The primary objective of this research is to advance subgovernment theory. As McCool (1989, 1990, 1995, 1998) has noted, "The phenomena of fragmentation and accommodation, and related concepts, are common themes in much of the literature on subgovernments and their principle participants" (McCool, 1989, p. 266). While much has been written concerning the expansion of competition and activity within subgovernments, the literature has not explored a number of external influences on subgovernments (Klyza, 1996). McCool (1989) suggests a need for political science to address four sets of questions in order to fully understand the role of subgovernments in contemporary policymaking. These include: (1) "what are the factors that affect the relative power of subgovernment participants, (2) what are the conditions and factors that provoke change in subgovernments, and (3) what variables affect the level of integration between subgovernments and their external environment, and (4) what are the democratic implications of subgovernments" (pp. 280–281)?

McCool (1998) also argues that we know little about how subgovernments behave during periods of conflict and proposes a framework through which researchers might identify the functional characteristics of subgovernments during periods of political conflict. McCool's (1998) "hierarchy of conflict" framework "permits the development of a typology of conflict, and an association between types of conflict and strategies" (p. 562). As McCool (1998) notes, shifting the research emphasis of subgovernments away from structure and toward identifiable behavior "provides a new definition of a sub[government]" as well as advancing the construction of sub-government typologies and the probability of strategic responses to differing kinds of conflict. McCool (1998) offers three testable hypotheses that might "yield insight into the causal relationships between elements of the political context, and the strategies employed by various kinds of sub[governments]" (pp. 565–566).[6]

---

[6] Note: *H1*: As the relative power of opposing sub[governments] in a policy conflict approaches equity, the greater the probability that conflict will move up the hierarchy...the higher the conflict moves up the hierarchy, the greater the probability that sub[governments] will operate in the strategic context of pluralized or conflictual sub[governments]...Conversely, the lower the conflict on the hierarchy, the greater the probability that sub[governments] will operate in the strategic context of autonomous or dominant/dissident sub[governments].

H2: The more government largess is perceived zero-sum, the greater the probability that sub[governments] will operate in the strategic context of dominant/dissident or conflictual sub[governments]. Conversely, the more government largess is perceived as non-zero-sum, the greater the probability that sub[governments] will operate in the strategic context of autonomous or pluralized sub[governments].

H3: As policymaking moves further down the conflict hierarchy, the greater the probability that the stronger sub[government] will protect the status quo. Conversely, as policymaking moves up the hierarchy, the greater the probability that all sub[governments] will work to alter the status quo in an effort to gain competitive advantage. From this it follows that if one subsystem is more successful improving its competitive advantage, then policymaking will start moving back down the conflict hierarchy.

Based on McCool's (1989, 1990, 1995, 1998) analysis, two things are clear: first, the essential questions concerning the dynamics of subgovernments when external factors foster change remain unanswered and, second, the essential questions concerning the dynamics of subgovernments when internal factors foster change remain unanswered. Therefore, in order to advance subgovernment theory, both the external and the internal factors that foster change must be addressed during a period of political upheaval.

This research *identifies conditions* under which subgovernment actors *strategically respond* to a *political conflict*. In doing this, McCool's (1989) four sets of questions are addressed and empirical evidence is provided in support of McCool's (1998) analytical framework of a hierarchy of conflict to "improve the validity and usefulness of the sub[government] model" (p. 566). The theory is advanced by explaining how subgovernments behave during times of significant political conflict and how they are affected by changes in presidential administrations.

A Case Study Exploring Stakeholder Conflict

The research within this book is a case study that is the result of taking a mixed methodological approach. This approach is a necessary one because of the complex dynamics concerning how political change and action within an existing body of laws led to the BLM's shift away from a rancher-dominated agency and toward an energy-dominated agency that remains to this day. While the term "case study" has multiple meanings, this is an "instrumental case study" because the research is intended "mainly to provide insight into an issue or to redraw a generalization" (Stake, 2003, p. 445). The BLM itself is of secondary interest but provides the setting necessary for an examination of changes in subgovernments due to external factors and internal conflict. The BLM has a historically robust subgovernment system that has been examined by a number of political scientists to further their own theories of subgovernments.

The expansion of domestic energy production at the direction of the president, vice president, and administrative appointees provides the setting for disrupting the relative stability of the BLM's land-use subgovernment. The expansion of domestic energy development challenged the deference to ranching interests that long typified the BLM's land-use policy. As noted earlier, researchers have been critical of the BLM because of how the BLM has been dominated by ranching interests since the agency's creation. More recent research indicates that the BLM's land-use subgovernment has been more inclusive of other land-use interests in shaping land-use policy. This same research also concludes that other land-use interests and government policymakers remain deferential to ranching interests regardless of any legislative alterations to the BLM's land-use policies. While previous findings suggest significant political tension between interest groups, the general conclusion remains that ranching interests are the politically dominant force in the BLM's land-use policymaking subgovernment. In many respects, the assumption that ranching is dominant over BLM land-use policy has limited the advancement of subgovernment theory.[7]

Previous subgovernment research has also focused on the dynamics between conflicting land-use policy interests. Primarily, this has meant research focused on environmental protection interests opposing ranching and other large resource use-related interests. The research written of here investigates a significantly different case. The focus of the exploration and analysis is the condition and dynamics of a political conflict that emerges between formerly allied—not opposing—land-use interests: energy and ranching. While previous research is correct in concluding that influence by environmental protection interests helps shape the BLM's land-use policymaking, the research also concludes that the BLM remains deferential to ranching interests. Previous research has assumed that ranching interests were so politically powerful and influential within the context of the BLM's land-use subgovernment that even the energy development interests deferred to and accommodated ranching.

The central thesis is that a change in presidential administration can lead to a shift in the domination of the BLM's land-use subgovernment. While multiple indicators suggest that a shift has now occurred, this research effort clearly demonstrates that the BLM's land-use policymaking now defers to, or accommodates, energy development interests. Finally, by demonstrating how political control of a policy subgovernment long dominated by one interest group has been wrested away by a formerly allied interest group, the research contributes to subgovernment theory.

In developing the argument, the research explores a significant chain of events that resulted in the disruption of control for an established, relatively stable subgovernment. In articulating the conditions for annexation of the subgovernment, the research (1) addresses how the actions of the executive branch awakened long-dormant statutes (Stock-Raising Homestead Act of 1916, §291 et seq; Mineral Leasing Act of 1920, §§2319–2328, 2331, 2333 2338, 2344; General Mining Act of 1872, §91) that established the legal dominance of energy development, which was, in turn, (2) manifested in the triggering mechanism of split-estate energy development and the strategic, competitive actions of ranching and energy interests that (3) resulted in a shift in the domination of the BLM's land-use policymaking subgovernment. The research documents how this causal process, in which an external change triggered a mechanism, led to policy change and conflict that resulted in energy development interests gaining control of the BLM's land-use policymaking subgovernment.

Analysis of the case study uses process tracing, a two-pronged approach of analysis (George & Bennett, 2005). First, archival materials are used to provide indicators of the conflict as it manifested over time and the actors, organizations, and political institutions involved. Second, the archival document analysis is supported by data drawn directly from elite actor interviews to confirm the analysis of histori-

---

[7]Note: The limited advancement of subgovernment theory is not necessarily limited to studies concerning the BLM but that the BLM and its subgovernment are often used as examples for discussion of subgovernment theory. Simply stated the BLM's subgovernment is a much studied subgovernment, but is hardly the only setting in which subgovernments are explored.

cal, governmental, and journalistic documents that identify the conditions under which the shift in policy dominance within the BLM occurred.

Thus, the book provides an analysis of the following archival material and historical secondary sources: (1) executive and legislative documents directing the BLM to administer the sale and management of energy development leases on both public and private lands; (2) primary government documentation of the impact of executive branch actions on the BLM's capacity to administer the energy leasing process and the consequences for policy favoring energy development, especially split-estate energy development; (3) news sources and journalistic accounts reporting ranching interest responses to energy development's legal domination and split-estate energy development and photographic representation of the impact of expanded energy leasing and development on private surface land; (4) archival documents and elite interviews documenting ranching interests' strategy of seeking state legislative protection of their surface lands from energy development; (5) archival documents and elite interviews documenting energy industry's strategic responses to ranching's legislative initiatives; (6) archival documents and elite interviews documenting state legislative responses to both ranching and energy's strategic legislative actions; and (7) archival documents and elite interviews documenting BLM's response to state legislation enacted to protect ranching interests from energy development of the federally managed mineral estate.

Finally, the book accounts for evidence identifying and comparing the conditions under which the conflict occurred across the states of New Mexico, Colorado, and Wyoming. These states are where the conflict between ranching and energy development interests first emerged and where these powerful interests have most openly competed for domination of the BLM's land-use policymaking subgovernment.

These findings are supported with a series of elite interviews of actors closest to the conflict in those three states. In other words, it is insufficient to present archival evidence that merely hints at the existence of conditions, causal chains of events, and triggering mechanisms to a political conflict. A comprehensive causal explanation becomes clearer by combining an analysis of documentary evidence with firsthand accounts from elite participants closest to the events and outcomes in question.

Interviewing small numbers of executive decision-makers, rather than a larger sample of participants, supports the process tracing methodological approach by confirming the initial analysis of a very specific series of events or processes. Because initial evidence merely suggests that ranching's dominance of the BLM's land-use subgovernment has subsided, support for those findings is evidenced by interviewing small groups of executive decision-makers. These interviews are necessary because initial findings indicate that most strategic decisions in the political conflict under investigation here were made by elites within a rather small set of groups across similar settings. The conflict between ranching and energy interests is, however, still evolving. Although this research effort had a definitive starting point, the conflict it analyzes is unlikely to end anytime soon.

The interviews presented within the book were conducted with a carefully selected set of elite decision-makers and collected firsthand accounts regarding the critical conditions, events, and mechanisms as well as the strategically competitive political relationships that exist among the actors. By virtue of their positions and proximity to the competition for dominance of the BLM's land-use subgovernment, interviews of senior actors—decision-makers—are essential for reconstructing the events and processes that are of interest (Tansey, 2007).

The criterion for selecting elites to interview was guided by two sampling methods: first, "purposive sampling" where knowledge of the process and actors under investigation guides the identification of the most appropriate individuals of interest and, second, "snowball or chain referral sampling" where the intimate knowledge among the principal actors of other, often more influential actors is known only to themselves and must be identified by the researcher's asking the question: "Who else should I be speaking to" (Tansey, 2007)?

Elite interview subjects were then selected based on their specific position and occupation within a small set of identifiable group(s) who were central to the conflict and competition for control of the BLM's land-use subgovernment. Second, others were selected for interview based on those initially interviewed who referred to others known to them as being influential, but who acted behind the scenes. This combination proved extremely useful and informative. Members of three groups, in the three states, were identified as the most appropriate interview subjects. The three groups are the following: (1) government officials including federal DOI appointees and state BLM administrators with decision-making authority concerning the management, resources, and oversight of ranching and energy development, as state legislators were engaged with stakeholder groups battling over passage of state surface owner protection acts; (2) state petroleum associations actively opposing the efforts of ranching supported interest groups sponsoring those state surface protection acts; and (3) ranching supported interest groups actively sponsoring and supporting those state surface owner protection acts.

I was not disappointed with the data collected from these interviews, and they are the heart of the analysis within this book. As expected, elite interviews of similarly situated individuals among similar groups in all three states confirmed and supported findings of conditions, causal events, mechanisms, political conflicts, and strategic competition from my archival research and analysis.

Conclusion

In conducting the research, six significant contributions to the field of political science are made. First, conditions and factors that affect the relative power within a "strong corner" of allied subgovernment participants are identified. Second, how presidential action may create competition and change in the control of subgovernments is accounted for. Third, variables affecting the level of integration between subgovernments and their internal and external environments are documented. Fourth, a more specific understanding of a very powerful subgovernment, its capacity for change, and the democratic implications of that change is developed. Fifth, a contribution is made to an existing body of Western land-use literature by providing a contextually deep analysis of a causal process where the unintended consequence of executive action released a causal mechanism—split-estate energy development of Western ranching lands—triggering political upheaval, conflict, and competition between Western landowners, energy development companies, and government. Finally, the analysis demonstrates that resulting from a politically induced upheaval to the BLM's land-use subgovernment, a shift in control over the BLM away from a rancher-dominated subgovernment to an energy development-dominated subgovernment has occurred.

Chapters

Chapter 2 includes a review of previous literature and documentation to trace the historic development of federal grazing and energy development legislation and policies. It includes an analysis of the historical legislative record to demonstrate how the federal government's desire to manage the use of public lands during the late nineteenth and early twentieth centuries helped establish the subgovernment relationships between the federal government, ranching interests, and the energy industry. Third, the chapter includes a description of the developing relationships and dormant legal conditions that enabled the modern struggle for dominance of the BLM's land-use subgovernment between the formerly allied interests of ranching and energy development.

Chapter 3 presents analysis of how the Bush-era executive branch actions altered federal domestic energy policies and have affected the BLM's domestic energy policies and its resource allocation. In this chapter archival and government documents describing executive branch actions that directed the BLM to favor the energy development industry are analyzed. The events and actions are presented chronologically to illustrate how the president and his executive appointees established, possibly unintentionally, the conditions for an impending political conflict. These events and actions resulted in increased levels, numbers, and types of federal energy development projects in New Mexico, Colorado, and Wyoming. Chapter 3 concludes with a brief analysis suggesting that increased split-estate energy development, under long-dormant legislation, triggered conflict and competition between the formerly allied interests of ranching and energy development.

Chapter 4 contains data gathered from primary government documents and journalistic sources across three settings, New Mexico, Colorado, and Wyoming, where a clear majority of split-estate energy development and conflict between ranching and energy development interests occurred. In this chapter changes in federal domestic energy policy resulted in increased split-estate energy development and the effects of that development are discussed. This analysis demonstrates that as split-estate energy development increased, the level of frustration among ranchers with the BLM also increased. In this chapter the spiraling conflict is documented from the evidence demonstrating that ranching organizations engaged in petitioning their state legislatures for protection from energy development. As ranching interests turned to state legislatures for protection, these accounts describe how and why the formerly allied interests of ranching and energy developers increasingly competed for control of the BLM's land-use subgovernment.

Chapters 5, 6 and 7 contain analysis drawn from the firsthand accounts among the decision-making elites gathered from the three primary groups who engaged in the conflict: (1) government officials, (2) state petroleum associations, and (3) ranching supported interest groups. The narrative analysis presented in Chaps. 5, 6 and 7 reinforce the previous chapters' documentary analysis of how the changes to domestic energy policy affected the internal stability of these established subgovernment actors. The method of corroboration employs firsthand accounts and responses to open-ended questions as an opportunity to clarify the impact executive decision-making had on the BLM's policymaking environment and strongly support the president's potential for disrupting established subgovernment alliances. The accounts of elites also lend a confirming voice to the conditions under which strategic competitive behavior for dominance of the subgovernment occurs. Additionally, the chapter contains an exploration of how external political upheaval and internal strife affect interest groups and their relationship with other established subgovernment actors. Finally, the chapter concludes with a comparative analysis of similarities and differences in the conditions, events, triggering mechanisms, political strategies, and competition voiced by those elites who had directly engaged in the political battles in Wyoming, New Mexico, and Colorado. The analysis presents these findings as a means of contextualizing the political conflict and providing deeper understanding to the theoretical concept of subgovernment domination in policymaking.

Chapter 8 concluded with a review of the evidence and analyses presented in previous chapters, reaffirming the central argument: The BLM has shifted from a rancher-dominated agency to an energy-dominated agency. Moreover, the evidence is once again discussed to illustrate to the reader that a change in policy initiated by the executive branch can trigger an unintended consequence that can then establish conditions for a political upheaval, conflict, and competition among allied interests within a policy subgovernment. These findings support the point that annexation of an agency's policymaking subgovernment is possible. In this chapter it is hoped that these findings, collectively, advance our understanding of subgovernment theory by (a) establishing the extent of presidential control over entrenched subgovernments, (b) providing implications for future political appointees and the land management agencies they direct, (c) mapping the strategic behavior of subgovernment actors during periods of political conflict, and (d) discussing the democratic implications of energy development's annexation of the BLM's subgovernment. Finally, the chapter presents a brief discussion of why the unforeseen shift in domination of a resource management agency's subgovernment by another large resource interest presents a host of unanswered questions that beg future research.

References

Baumgartner, F. R., & Jones, B. D. (1993). *Agendas and instability in American politics*. Chicago: University of Chicago Press.

Browne, W. P. (1988). *Private interests, public policy, and American agriculture*. Lawrence, KS: University Press of Kansas.

Bureau of Land Management. (n.d.). Split estate. Retrieved August 2006, from http://www.blm.gov/bmp/Split_Estate.htm

Cahn, M. A. (1995). *Environmental deceptions: The tension between liberalism and environmental policymaking in the United States*. Albany, NY: State University of New York Press.

Castelnuovo, R. (1998). Turning NEPA on its head: Assessments that advance property rights at the expense of the environment. In H. M. Jacobs (Ed.), *Who owns America?: Social conflict over property rights* (pp. 54–78). Madison, WI: The University of Wisconsin Press.

Cater, D. (1964). *Power in Washington*. New York: Random House.

Cawley, R. M. (1993). *Federal land western anger: The sagebrush rebellion and environmental politics*. Lawrence, KS: University of Kansas Press.

Clarke, J. N., & McCool, D. C. (1996). *Staking out the terrain: Power and performance among natural resource agencies* (2nd ed.). Albany, NY: State University of New York Press.

Clary, D. (1986). *Timber and the forest service*. Lawrence, KS: University Press of Kansas.

Culhane, P. J. (1981). *Public lands politics: Interest group influence on the forest service and the bureau of land management*. Baltimore: Johns Hopkins University Press. for Resources for the Future.

Dahl, R. A. (1967). *Pluralist democracy in the United States: Conflict and consent*. Chicago: Rand McNally.

Davis, C. (1997a). Politics and public rangeland policy. In C. Davis (Ed.), *Western public lands and environmental politics* (pp. 87–110). Boulder, CO: Westview Press.

Davis, D. H. (1997b). Energy on federal lands. In C. Davis (Ed.), *Western public lands and environmental politics* (pp. 141–168). Boulder, CO: Westview Press.

Donahue, D. L. (1999). *The western range revisited: Removing livestock from public lands to conserve native biodiversity*. Norman, OK: University of Oklahoma Press.

Durant, R. F. (1992). *The administrative presidency revisited: Public lands, the BLM, and the Reagan revolution*. Albany, NY: State University of New York Press.

Earthworks. (n.d.). Surface owner protection legislation. Retrieved from http://www.earthworks-action.org/SOPLegislation.cfm

Eilperin, J. (2006, July 25). Growing coalition opposes drilling: In N.M. battle, hunters team with environmentalists. *The Washington Post*. Retrieved from http://www.washingtonpost.com/wp-dyn/content/article/2006/07/24/AR2006072400951.html

Energy Policy Act of 2005 (Title 42 of the *United States Code*).

Environmental Working Group. (2004). Who owns the West? Oil and gas leases. Retrieved from http://www.ewg.org/oil_and_gas/

Foss, P. O. (1960). *Politics and grass: The administration of grazing on the public domain*. Seattle, WA: University of Washington Press.

Freeman, J. L. (1965). *The political process* (2nd ed.). New York: Random House.

General Mining Act of 1872 (Title 30 of the *United States Code*, §§ 22–54).

George, A. L., & Bennett, A. (2005). *Case studies and theory development in the social sciences*. Cambridge, MA: MIT Press.

Hage, W. (1994). *Storm over rangelands: Private rights in federal lands* (3rd ed.). Bellevue, WA: Free Enterprise Press.

Hardin, B., & Jehl, D. (2002, December 29). Ranchers bristle as gas wells loom on the range. *New York Times*. Retrieved from http://www.nytimes.com/2002/12/29national/29METH.html

Heclo, H. (1978). Issue networks and the executive establishment. In A. King (Ed.), *The new American political system* (pp. 95–105). Washington DC: American Enterprise Institute.

Ingram, H. (1990). *Water politics*. Albuquerque, NM: University of New Mexico Press.

Jones, C. O., & Strahan, R. (1985, May). The effect of energy politics on congressional and executive organizations in the 1970s. *Legislative Studies Quarterly, 10*, 151–178.

Kelso, W. (1995). Three types of pluralism. In D. C. McCool (Ed.), *Public policy theories, models and concepts: An anthology* (pp. 41–54). Englewood Cliffs, NJ: Prentice Hall.

Kingdon, J. (1984). *Agendas, alternatives, and public policies*. Boston: Little, Brown.

Klyza, C. M. (1996). *Who controls public lands?: Mining, forestry, and grazing politics 1870– 1990*. Chapel Hill, NC: The University of North Carolina Press.

Knight, R. L., Gilgert, W. C., & Marston, E. (Eds.). (2002). *Ranching west of the 100ᵗʰ meridian: Culture, ecology, and economics*. Washington, DC: Island Press.

Lofholm, N., & McGuire, K. (2006, July 2). Boom life: As energy drilling explodes across the West, those in its path try to keep, or adjust, their ways amid the upheaval. *The Denver Post*. Retrieved from www.denverpost.com/search/ci_4003872

Lowi, T. (1979). *The end of liberalism* (2nd ed.). New York: W.W. Norton.

McConnell, G. (1966). *Private power and American democracy*. New York: Random House.

McCool, D. C. (1987). Command of the waters. Berkeley, CA: University of California Press.

McCool, D. C. (1989). Subgovernments and the impact of policy fragmentation and accommodation. *Policy Studies Review, 8*(4), 264–287. https://doi.org/10.1111/j.15411338.1988.tb01101.x

McCool, D. C. (1990, Summer). Subgovernments as determinants of political viability. *Political Science Quarterly, 105*(2), 269–293.

McCool, D. C. (1995). *Public policy theories, models, and concepts: An anthology*. Englewood Cliffs, NJ: Prentice Hall.

McCool, D. C. (1998, June). The subsystem family of concepts: A critique and proposal. *Political Science Quarterly, 51*(2), 551–570.

Merrill, K. R. (2002). *Public lands and political meaning: Ranchers, the government, and the property between them*. Berkeley, CA: University of California Press.

Miller, A. C., Hamburger, T., & Cart, J. (2004, August 25). White House puts the west on fast track for oil, gas drilling. *Los Angeles Times*. Retrieved from www.latimes.com/news/yahoo/la-na-bog25aug25,1,500016.story

Mineral Leasing Act of 1920, as amended (Title 30 of the *United States Code* § 181 et seq.).

Nie, M. (2008). *The governance of western public lands*. Lawrence, KS: University Press of Kansas.

Rosenbaum, W. (1993). Energy politics in the West. In Z. Smith (Ed.), *Environmental politics and policy in the West*. Dubuque, IA: Kendall/Hunt.

Sabatier, P. A., & Jenkins-Smith, H. C. (1999). *Theories of the policy process*. Boulder, CO: Westview Press.

Setback for drilling agreement [Editorial]. (2006, April 15). The Denver Post. p. C-11. Retrieved from https://www.denverpost.com/2006/04/14/setback-for-drilling-agreement/

Smith, Z. A., & Freemuth, J. C. (Eds.). (2007). Environmental politics and policy in the west (Revised ed.). Boulder, CO: University Press of Colorado.

Stake, R. E. (2003). Case studies. In N. K. Denzin & Y. S. Lincoln (Eds.), *Strategies of qualitative inquiry* (2nd ed.). Thousand Oaks, CA: Sage Publications.

Starrs, P. F. (1998). *Let the cowboy ride: Cattle ranching in the American west*. Baltimore: The Johns Hopkins University Press.

Stock Raising Homestead Act of 1916 (Title 30 of the *United States Code*).

Tansey, O. (2007, October). Process tracing and elite interviewing: A case for non probability sampling. *Political Science and Politics, 40*(4), 765–772.

The Wilderness Society. (n.d.). Abuse of trust: A brief history of the Bush administration's disastrous oil and gas development policies in the Rocky Mountain West. Retrieved from http://www.wilderness.org/Library/Documents/AbuseOfTrust.cfm

Truman, D. (1971). *The governmental process: Political interests and public opinion*. New York: Knopf Press.

Western Organization of Resource Councils. (n.d.). Supporting declarations. Oil and gas industry responsibility petition. Retrieved from http://www.worc.org/issues/art_issues/bonding_petition.html

Wilkinson, C. F. (1992). *Crossing the next meridian: Land, water, and the future of the West.* Washington, DC: Island Press.

Wilkinson, T. (2005, May 10). Energy boom is crowding ranchers: More ranchers rail against federal 'split estate' laws that control mineral rights beneath their land. *The Christian Science Monitor.* Retrieved from www.csmonitor.com/2005/0510/p01s02-usju.html

Yandle, B. (Ed.). (1995). *Land rights: The 1990s' property rights rebellion.* Lanham, MD: Rowman & Littlefield Publishers, Inc.

Chapter 2
Legal History

Abstract Analysis of the historical record demonstrates how the federal government's desire to manage the use of public lands during the late nineteenth and early twentieth centuries helped establish the dynamic and always evolving subgovernment relationships between the federal government, ranchers, and energy developers. It is in the description of this context and these dynamics that political observers can discern how these developed relationships and dormant legal conditions have come to enable the modern struggle for dominance of the BLM's land-use subgovernment. The legislative and legal manner these interests have, in almost parallel fashion, developed over time suggests that the clashing of interests within the land-use subgovernment of the BLM was inevitable.

Keywords Bureau of Land Management · Federal law · Ranching · Energy · Public policy · Interest groups · Federal legislation

This chapter reviews previous literature and documentation tracing the historical development of federal grazing and energy development legislation and policies. The chapter includes a preliminary analysis using the historical record to demonstrate how the federal government's desire to manage the use of public lands during the late nineteenth and early twentieth centuries helped establish the subgovernment relationships between the federal government, ranching interests, and the energy industry. The chapter also includes a description of the developing relationships and dormant legal conditions that eventually help enable the modern struggle for dominance between the formerly allied interests of ranching and energy development of the Bureau of Land Management's (BLM) land-use subgovernment.

History of Federal Land Management: Ranching

For the first 200 years of American history, federal land policy centered on facilitating the sale or transfer of public lands to states or private parties for the benefit of the nation. The retention and management of public lands and their resources by

© Springer Nature Switzerland AG 2019
R. E. Forbis Jr., *Altered Policy Landscapes*,
https://doi.org/10.1007/978-3-030-04774-0_2

federal agencies is a relatively modern phenomenon. For instance, until 1976, federal land policy still generally favored the disposition of public lands and resources to private interests in order that the nation is "tamed, farmed, and developed" (Coggins & Wilkinson, 1987, p. 47; Wilkinson, 1992. p. 34).

Federal land management policy can ultimately be traced to debates between Alexander Hamilton, who advocated the sale of public lands to assist in the effort to pay Revolutionary War debts, and Thomas Jefferson, who advocated the fostering and development of an agrarian-based society by providing free land to frontier settlers (Anderson & Hill, 1990). Three general public land disposal policies emerged from these debates: (1) sale of the land to the highest bidder with no required residence or occupation of the land; (2) title by first occupancy of the land, or giving squatters first right of purchase at a minimum price; and (3) title to the land by homesteading, or providing free or minimally priced land to any settler who satisfied specific requirements of residency and/or improvements. Each of these land disposal policies emerged as a result of differing opinions and compromise, regarding the fundamental debate of how best to settle the Western frontier of an emerging nation. Each of these policies will be discussed in turn.

For the first 50 years after the Revolutionary War, land policy was dominated by the Hamiltonian position that the best interest of the nation was served by a land management policy that encouraged Western expansion by selling public lands to the highest bidder. In turn, Hamilton also believed that the sale of public lands would generate sufficient funds to satisfy the debt incurred by the war. The nation's two primary sources of revenue at the time were tariffs on imported goods and the sale of public lands.

The Land Act of 1797 was passed to facilitate the sale of public lands by means of a survey system based on an auction system of 36 section rectangular units at a starting minimum bid of $2.00 per acre. During this same period of time, land was sold to individuals and companies acting as brokers for small groups or individuals seeking to purchase land. Prices and terms of ownership were relatively uncomplicated as the details were negotiated between those acting for the buyers of land and Congress. Notably, during this period, railroad companies—the industrial mechanism of westward expansion—purchased or were granted lands directly from the federal government. By 1812 land sales and purchases had reached a point where more formal administration of the program was deemed necessary. As a result, Congress created the General Land Office. This first federal land management office would, in 1946, be absorbed by the BLM (Coggins & Wilkinson, 1987, p.106–128; Merrill, 2002).

Squatting was a common practice in America in both the pre- and post-Revolutionary War eras. The practice created tension between the federal government's desire to accommodate settlement of the Western frontier and its interest in protecting the lawful purchase and legal title to lands of the Western frontier. According to Thomas Jefferson in 1776, "they will settle the lands in spite of everybody" (Coggins & Wilkinson, 1987, p. 88). The federal government's frustration with the practice of squatting is exemplified by the instances in which federal troops were sent to dispatch squatters from land that was legally owned by either the federal government or private speculators of federal lands (Merrill, 2002).

In 1805, with the national debt retired and more of the American population in Western states expanding, pressure to accommodate squatters and the practice of squatting on public lands grew. Thus, the federal government adopted a system known simply as "preemption to squatters." This land-use policy recognized the right of property as a result of labor or improvement to the land by occupation of the land. Through a series of legislative acts in the first half of the nineteenth century, the federal government implemented a policy that gave the preferential right of settlers-squatters to buy their claims at modest prices without competitive bidding. The policy of preemption to squatters continued as a retroactive response to the practice of squatting on public lands until 1841.

In 1841, Congress passed the General Preemption Act, which reformed the retroactive policy of preemption to squatters, to a prospective policy of preemption to squatters only on lands previously surveyed by the federal government. This reform was the result of many abuses of the retroactive nature of the pre-1841 land management policy that gave preemption to squatters. Many abuses were reported at the time of the legislation's passing, especially reports of fraud and the somewhat common practice of squatters occupying timberland just long enough to strip it bare of its resources only to move on to a new claim. Some 50 years later, the preemption to squatters, regardless of its retroactive or prospective nature, was deemed a policy badly in need of reform or abolishment. In 1891, Congress repealed the Preemption Act of 1841, but the General Land Office and its administration to the management of public lands remained in place.

At the close of the Civil War, two developments—one economic in nature and the other social in nature—initiated a major reform in federal land management policy: (1) revenues from the sale of public lands as a percentage of the total federal budget had begun declining sharply, and (2) emigration and industrialization had created a class of landless and unemployed workers in the Eastern United States. Speaking to the second development, and arguing forcefully for a system of homesteading in the West, a representative from Illinois remarked:

> Unless the government shall grant head rights…prairies, with their gorgeous growth of flowers, their green carpeting, their lively lawns and gentle slopes, will for centuries continue to be the home of the wild deer and wolf; their stillness will be undisturbed by the jocund song of the farmer, and their deep and fertile soil unbroken by the ploughshare. Something must be done to remedy this evil. (Foss, 1960, p. 91)

In order to remedy "this evil," Congress passed the Homestead Act of 1862 authorizing "entry onto 160 acres of any land subject to preemption, later extended to unsurveyed lands where Indian title was extinguished" (Preemption Act of 1862, ch. 94, § 1, 12 stat 413). This statute governed the disposal of public lands for the very Jeffersonian land management policy of providing public lands for free or at a minimal price for the purpose of fostering and developing an agrarian-based society of frontier settlers. Indeed, the concept of homesteading would guide the federal policy for management of public lands and resources for the next 114 years (Coggins & Wilkinson, 1987, p. 91; Merrill, 2002). Land acquired under the statute was free, except for filing fees. The sole requirement of the settler was to occupy the land for 6 months in order to establish clear residency after application had been made to the General Land Management Office. By the 1880s homesteading on the American

Western frontier constituted the majority of new farms in the United States, and between 1868 and 1904, nearly 100 million acres of land were homesteaded by pioneers (Coggins & Wilkinson, 1987, p. 92).

As Karen R. Merrill (2002) points out, while homesteading legislation was enacted for the expressed purpose of expanding the agrarian vision of a nation of working farms and families, legislation was also enacted as a policy solution for controlling the unregulated grazing practices of "King Cattle." Known as the "Texas System," the practice of open range grazing is rooted in herding traditions which dates to the close of the eighteenth century when cattle and cowboys of the Carolinas and Mexico roamed vast areas of open space (Merrill, 2002, p. 18). Hence, grazing on the Western frontier lands of the United States was governed by a series of unwritten codes and informal laws by ranchers who viewed the open range as a free resource for their use. In turn, free use of the open range, unregulated by government authority, established the conditions for the expansion of the Western ranching industry. Overgrazing was problematic, and the large cattle ranchers of the West organized themselves in such a manner as to create almost government-like institutions for the purpose of controlling grazing on the open range.

Undisturbed by government authority, ranchers created a rather simple system for governing and enforcing the unwritten codes of behavior. For example, in order to maintain control of grazing districts and account for their property (e.g., cattle) at the time of roundups, ranchers devised a system of brand identification. These brands were registered with stock growers associations, which were organized and managed by the relatively few large barons of the cattle industry in the West. Unregistered brands, or the alteration of brands, unapproved fencing of the open range, or the trespass onto the open range by any stock other than branded cattle were considered offenses and enforceable by the now infamous "Code of the West." In fact, many of these conflicts were settled by "hired regulators" who, under the direction of the cattle baron dominated stock growers associations, killed suspected offenders and engaged in systematic harassment of farmers and sheepherders taking advantage of federal homesteading laws that were designed to encourage the agrarian settlement of the Western range (Merrill, 2002, pp. 23–29). The silence of the Congress on the land-use customs and codes that had grown out of these grazing practices "was taken to be its most important policy statement, for that silence clearly translated into permission" (Merrill, 2002, p. 26). Commenting on the customs, traditions, and implied policy of unfettered use, governing the use of the open range for cattle grazing, the Supreme Court noted in *Buford v. Houtz* (1890) that:

> There is an implied license, growing out of the custom of nearly a hundred years, that the public lands of the United States…shall be free to the people who seek to use them where they are left open and unenclosed, and no act of government forbids this use…The government of the United States, in all its branches, has known of this use, has never forbidden it, not taken any steps to arrest it. (Merrill, 2002, p. 26)

Arguments that spurred passage of homestead legislation by Congress were then motivated by two factors. First, Congress was motivated by a desire to establish conditions that would encourage the creation of a Western society of citizen-farmers. Second, Congress and federal land managers were motivated by a desire to break the grip of the relatively few large cattle barons who had taken control of the West's open range. While homesteading legislation proved successful in breaking the grip of the cattle barons and their use of the open range, nonetheless, it was during the era of homesteading that ranching and farming became effectively intertwined with each other (Merrill, 2002). Simply put, grazing the land and tilling the land became non-distinct in the language of agriculture as well as federal land-use management policy.

Even though the era of the federal government's free frontier policy had come to an end by 1890, Congress remained intent on further disposing itself of federally owned lands. For example, by 1903 almost 9 million acres of land in the State of Nebraska had either not been claimed or had been abandoned. The most common claim for abandoning previously claimed homestead lands was that the soil was not arable. In the example referenced above, this meant that in one area of Nebraska homestead lands, 247 of the 250 claims were abandoned for being deemed unsuitable for agriculture. And, it was not uncommon practice for ranching operations to stake claims of property using federal homestead legislation. Once homesteading became available, ranchers took advantage of the generous public land laws to "gain control of their public ranges" (Merrill, 2002, p. 27). Even though the practice of acquiring open range by ranchers did not achieve the envisioned agrarian social unit Congress had hoped for (Gates, 1968, pp. 435–62; Friedman, 1985, pp. 416–421), Congress remained undeterred and passed the Enlarged Homestead Act of 1909 and the Stock-Raising Homestead Act of 1916 which extended, enlarged, and established the land management policies associated with homesteading for the raising of livestock or crops. These two homesteading acts authorized homestead entry to 640 acres and 320 acres, respectively (Coggins & Wilkinson, 1987, p 96).

And while these last two homestead acts "spurred the greatest run on homesteads since the passage of the Homestead Act" (Merrill, 2002, p. 43), it is significant that passage of the Stock-Raising Homestead Act of 1916 also reserved, and legally established, federal ownership of all the subsurface coal and mineral rights of all lands homesteaded per the 1916 act and thereafter. In effect, the Stock-Raising Homestead Act of 1916 legally severed the surface estate and its attendant property rights, from the subsurface estate and its attendant property rights. Significantly, the Stock-Raising Homestead Act of 1916 created an anomaly in American property law known simply as "split-estates."

History of Federal Land Management: Mining

Mining claims on federal lands are not governed by homesteading acts as discussed above. It should be recognized that, similar to cattle barons' grazing operations on the open range prior to the series of homesteading acts, mining camps of the Western frontier also operated under a set of complex unwritten laws and informal codes of behavior. Beginning with the discovery of gold at Sutter's Mill in 1848, and throughout 1849, an onslaught of gold-fevered treasure seekers converged upon California and the open lands of the American West, seeking both fame and fortune with the extraction of valuable minerals (Wilkinson, 1992, p. 34). Hundreds of mining camps, settlements, and collective encampments sprung up overnight serving as major forms of social units in the mineral regions of the West (Wilkinson, 1992, p. 38). Although mining camps often lay within federal and/or state jurisdiction, like the cattle barons grazing their cattle within close proximity of the mines, little if any formal law governed the practice of mining.

Fashioned out of necessity for structure and guidelines, the informal codes of mining camps governed mining during the period of discovery until 1866. Charles F. Wilkinson describes these informal codes as "…montages of Spanish rules transported north by Mexican miners, regulations from the Midwest, improvisation bred of commons sense, and local custom" (Wilkinson, 1992, pp. 38–39). The concept of "first in time, first in right" meant that the miner had exclusive right to that which the miner discovered. Importantly, these exclusive rights also included exclusive use of the resources available—most commonly water—for the purpose of extracting the mineral. This informal code of discovery and claim, as well as the attendant property rights, is an example of the unwritten rule of law frequently applied to miners and mining operations during this period of time (Wilkinson, 1992, p. 39).

The codes of the mining camps also helped establish procedures for the acquisition of individual mining claims and the extraction-related work performed on the claims. For example, in addition to being limited to a single claim, in order to establish a claim to a particular piece of property, in keeping with the informal nature of the mining codes, all a claimant had to do in order to "stake their claim" was to post some manner of written notice somewhere on the parcel of land or place a more permanent monument into the ground or mark the surrounding trees with a blaze or a brand. In turn, a duly elected recorder collected and maintained information relating to claims in a particular area or region. Successful claimants—those confident enough that their stake was properly vested and recorded with the proper authorizations—were required to maintain their claim by regularly performing mining-related work on their claims. Typically, this practice dictated that a claimant work their respective parcels of land at least 1 day per month during the course of the mining season (Wilkinson, 1992).

The informal codes of the mining camps even included provisions for settling disputes between competing claimants. Patterned after what is now commonly referred to as "dispute resolution," the two claimants would select an arbitrator who, in turn, selected an additional arbitrator to assist in the hearing and settling of the dispute. Arbitrators were typically required to be miners or residents of the nearest mining camp. Third-party outsiders, no matter their expertise, were frowned upon. Wilkinson notes that some camps actually prohibited lawyers from serving as arbitrators. For example, the Union District Court of Colorado famously declared that lawyers seeking to engage in the practice of arbitrating mining claims would suffer the punishment for engaging in such activities, declaring that, "[N]o lawyer shall be permitted to practice law in any court in the district under a penalty of not more than fifty nor less than twenty-five lashes and shall be forever banished from the district" (Gates, 1968, pp. 435–62; Friedman, 1985, pp. 416–421).

As Western territories gained statehood, the informal codes of the miners and mining camps, like the informal codes that had managed King Cattle's grazing of the open range, became the foundation for state and federal laws applicable to mining activities on both private and public lands. In California, for example, early state mining statutes proclaim that "the customs, usages or regulations established and in force at the bar or diggings embracing such claims...when not in conflict with the Constitution and laws of this State, shall govern the decision of the action" (Wilkinson, 1992, p. 39). Commenting on the legislative statute and noting the transition between the informal and the formal codification of mining practices, the California Supreme Court declared:

> These customs...were few, plain and simple, and well understood by those whom they originated...And it was wise policy on the part of the legislature not only to supplant them by legislative enactments, but on the contrary to give them the additional weight of a legislative sanction...Having received the sanction of the Legislature, they have become as much a part of the law of the land as the common law itself, which was not adopted in a more solemn form.... (Wilkinson, 1992, p. 39)

The federal government, unlike the attempt of mining states to engage in a more formal codification of mining customs, exercised little if any oversight of mining operations during this period of time, even though the majority portion of mining was taking place on federal lands and making use of federally owned resources in the process of extracting minerals (Wilkinson, 1992, p. 40). A small number of federal mining statutes did exist at the time, but these statutes only provided for the sale of federally owned mineral lands in the Eastern United States. With the exception of a very limited process of leasing federal lands for mining activity, which had been adopted in 1807 and then abandoned in 1846, no federal law governed mining activities in the Western United States until 1866. Thus, during this period of time, miners openly trespassed on federal lands, and mining practices were essentially left unregulated despite the federal government's implied, if not complicit, approval of mining activities (Brown as cited in Wilkinson, 1992, p. 38).

Like grazing on the open range, mining had become a central activity of American society in the Western United States well before 1866. An estimated 25,000 men worked the mines of California by the middle of the 1860s. Thirty percent of the population of states like Nevada and Idaho and as much as 25% of the population of Montana were working as miners during this period of time. Moreover, these figures do not reflect the number of individuals working to support the mining industry such as "...assayers, equipment manufacturers, teamsters, and suppliers of clothing, housing, and entertainment" (Brown as cited in Wilkinson, 1992, p. 38). In fact, during this period of time, the only area of the West not centrally dependent upon mining was the Utah territory, where farming still dominated the use of the lands and the activities of the communities.

The federal government's effort to regulate mining in the American West began with the Mining Act of 1866 (Brown as cited in Wilkinson, 1992).[1] The act declared

---

[1] Note: Mining Act of 1866 (repealed 1872). The official title of the 1866 Act was "An act granting the Right of Way to Ditch and Canal Owners over the Public Lands, and for other Purposes."

that "the mineral lands of the public domain, both surveyed and unsurveyed, are hereby declared to be free and open to exploration and occupation by all citizens…" (Wilkinson, 1992, p. 42). Although limited in scope to lode claims, covering only gold, silver, cinnabar, and copper, the 1866 Act established a zone encompassing close to an estimated billion acres of federal public lands for mining (Wilkinson, 1992, p. 42). Additionally, the act not only allowed a claimant to explore a mineral discovery "to any depth, with all its dips, variations and angles," it also granted claimants access to "a reasonable quantity of surface for the convenient working of the same" (Wilkinson, 1992, p. 252).

Despite the limited scope of the 1866 Act, it is important to note that, not unlike the General Preemption Act of 1841 that addressed the common practice of farming and ranching on the open range, the Act of 1866 represented the same implicit understanding that a powerless federal government recognized the legality of informal understandings and customs regarding the unregulated private use of public lands. For example, the 1866 Act provided mining would commence "subject…to the local customs or rules of miners in the several mining districts," so long as local customs or rules did not directly conflict with federal law (Wilkinson, 1992, p. 252). The bill's primary proponent, Senator William M. Stewart of Nevada, a former forty-niner and mining attorney, invoked images of the mining codes when he spoke in favor of the 1866 Act:

> The miner's law is part of the miner's nature. He made it. It is his own bantling, and he loves it, he trusts it, and obeys it. He is given the honest toil of life to discover wealth which when found is protected by no higher law than that enacted by himself under the implied sanction of a just and generous government (Wilkinson, 1992, p. 42). Because the act was so closely tied to the informal customs and codes of the mining camps of the American West, the 1866 act has been referred to as the Miner's Magna Carta. (Martz as cited in Wilkinson, 1992)

The general principles of the act were extended in 1870 to placer deposits and eventually became the basis for the General Mining Law of 1872, which survives largely intact to the present day (Wilkinson, 1992, p. 43).

The General Mining Law of 1872 was passed with the purpose of providing a solution to problems associated with the unregulated mining claims and practices in the American West. With its passage, Congress declared its policy to "promote the development of the mining resources of the United States" (General Mining Act of 1872 amended by the Mining and Minerals Policy Act of 1970).[2] And, while the Act of 1866 only addressed a small number of minerals, the 1872 Act provided mining access to "all valuable mineral deposits in lands belonging to the United States, both surveyed and unsurveyed." The act established that all mining activities for all mineral resources "shall be free and open to exploration and purchase, and the lands in which they are found to occupation and purchase…under regulations prescribed by law, and according to local customs or rules of the miners in the several mining districts, so far as they are…not inconsistent with the laws of the United States" (General Mining Act of 1872 (relevant to the 1872 Act))."

---

[2] Note: The official title of the 1872 Act is "An Act to Promote the Development of the Mining Resources of the United States."

At the time, "valuable" mineral deposits included "whatever is recognized as a mineral by the standard authorities on the subject" (Copp, 1882, as cited in Large, 1986. p. 50–51).[3] While the Act of 1872 still principally applies to hard-rock minerals, such as gold, silver, uranium, copper, iron, lead, aluminum, and gemstones, Congress amended this section and removed several types of minerals with the intent of providing separate legislation providing for their lease, extraction, development, and sale (Wilkinson, 1992, p. 44). These other minerals include resource commodities such as oil, gas, oil shale, and coal and other common materials including gravel, sand, and cinders (Wilkinson, 1992, p. 44).

Under the guidelines established by the General Mining Act of 1872, an individual can obtain both surface and subsurface mining rights to a particular parcel of land, called a "location" by the 1872 Act (General Mining Act of 1872, §26 (1994)). Even in the modern era, in order to establish a valid mining claim, a claimant must meet the following requirements: (1) a distinct mark must be placed on the "location" to be claimed, (2) the claim must be recorded at the local recorder's office, and (3) a fee of $100 must be paid annually to the federal government as means by which exclusive rights to the "location" are retained. If a claimant holds an unpatented claim, a patent can be secured for a fee as little as $5 per acre and submitting an annual statement certifying that a minimum of $500 worth of labor has been performed on the site (General Mining Act of 1872, §29 (1994)). And, when a patent is obtained by the claimant, a fee simple title to the "location" is transferred from the federal government to the individual (Graf, 1997).

Given the immense financial resources of modern extractive industries and their affiliate organizations, it might seem strange that the federal government still offers mining rights—and the rights of property that come with them—at such remarkably small prices. Viewed in the context of the historical development of mining laws, not unlike the fees for grazing allotments on federal lands, the government's economically grounded justifications for such generosity did make a bit of sense in the nineteenth century.

In the late nineteenth century, with Hamiltonian visions of America's Manifest Destiny and Jeffersonian dreams of a nations filled with yeoman farmers still dancing in the minds of federal legislators and other land-use policymakers, much of the public land-use policy relating to the American West remains centered on the federal government's effort to settle the great Western frontier west of the Mississippi River. As one commentator has expressed, these visions, dreams, and aspirations remain with us to the present day:

> Legislators viewed it as their duty to distribute the lands of the West to individual settlers, railroads, and entrepreneurs, thus implementing the American ideal. The permanent federal landholdings that are today part of the western landscape were not on the nineteenth-century agenda; the predominant view was that the federal government would eventually distribute all of its holdings to the states and individual settlers. (Knight, 2002, p. 626)

---

[3] Note: Defining "valuable" mineral deposits—A continuing quagmire

Legislative efforts regarding both homesteading and mining were efforts to populate the West and make the region economically viable while diminishing the problems associated with managing such large expanses of federally owned lands and resources. Invariably, these governmental efforts involved convincing settlers to venture into these dangerous and largely unknown areas. And really, what better way to accomplish the policy goals associated with Manifest Destiny and streamlining governmental management of problem areas than to entice an entire population with inexpensive access to lands suitable to agriculture and mining?

Dovetailed Federal Land Management: Ranching and Energy

Responsibility for grazing land policies was divided among an array of federal agencies—such as the National Forest Service and the Department of Agriculture—as the federal government sought policy solutions to the problem of overgrazing the open range of the American West. With passage of the Taylor Grazing Act of 1934, the expansive nature of the previously discussed homesteading and open range grazing eras of federal land management policy concluded (Donahue, 2005) that the act itself was instrumental in effectively reversing the homesteading and grazing policies of federal land management agencies. The Taylor Grazing Act is noteworthy because it sought to resolve the land management problem created by homesteading acts as well as the unregulated practice of grazing the open lands of the federal government. In doing so, the effect of the Taylor Grazing Act of 1934 was to blur the distinction between the activities of farming with the activity of raising and grazing cattle. It is also important to note that the Taylor Grazing Act also placed the management of all federal grazing lands under the direction of the Department of the Interior and, eventually, through the Bureau of Land Management.

Thus, it was in the early period of the twentieth century that the federal government began to develop and implement a wide variety of sustainable public land-use policies to offset the unsustainable private use of public lands and resources. For example, on the one hand, the Taylor Grazing Act of 1934s was passed with sustainability of the surface estate's open range in mind, while on the other hand the Mineral Lands Leasing Act of 1920[4] was passed with sustainability of the subsurface estate's sustainability in mind (Merrill, 2002). It is also important to note that the Mineral Lands Leasing Act also placed the management of all federal mineral leases under the direction of the Bureau of Land Management within the Department of the Interior. Like the previous responsibility for grazing land policies, previous mineral land policies had been divided among an array of federal agencies that included the National Forest Service and the Department of Agriculture.

In combination, these two acts of Congress, the Mineral Lands Leasing Act of 1920 and the Taylor Grazing Act of 1934, helped usher in the federal government's early twentieth-century shift away from the dispersal of the public domain benefiting

---

[4]Note: The Mineral Lands Lasing Act of 1920 is commonly referred to as the Mineral Leasing Act of 1920.

the individual and toward a land-use policy of sustainability benefiting the public. It is important to note that while the Taylor Grazing Act effectively repealed the practice of government sponsored homesteading, the act did not repeal the government's ownership of the subsurface minerals per the Livestock Raising Homestead Act of 1916. This appears a rather curious side note to history because both these acts were sponsored by Edward T. Taylor, who, at the time, was Colorado's senior congressman (Merrill, 2002). As mentioned previously, the conclusion of the federal government's free frontier policies gave way to the introduction and passage of government policies aimed at breaking King Cattle's control of the federally owned rangelands, as well as independent mining operators' (hereafter referred to as oil and gas, or simply as energy developers or industry) control of federally owned minerals. In essence, part of the appeal for passage of the Stock-Raising Homestead Act of 1916 was the severing of ownership and activities into separate, manageable, split-property estates.

The Mineral Lands Leasing Act of 1920 as well as the Taylor Grazing Act also establishes the federal government's claim of sovereign right of ownership to all remaining public lands and mineral resources. In addition to claiming proprietary decision-making rights over federal property, these acts targeted oil producers and cattle producers, respectively. These acts accomplished these goals by retaining the "nominal fee" element of the federal government's late nineteenth-century disbursement policies. The retention of the nominal fee elements of the late nineteenth-century legislation in the early twentieth-century legislation is reflected in both the subsurface estate's mineral lease policies of the Mineral Lands Leasing Act and the surface estate's grazing lease policies of the Taylor Grazing Act. In either case, leases must be secured from the managing federal agency—the BLM—before grazing or drilling permits are issued to the private user.

What this means is that the statutory requirement of securing grazing and mineral leases and/or permits are the regulatory tools by which the agency fulfills its legislative mandate to manage the public domain in the public's best interest. Additionally, the leasing and permitting process helps establish a contractual relationship between the property's owner—the federal government—and the lessee user of the property, the rancher or mineral developer. In turn, the leasing contract becomes a means by which the federal government's general policy goal of sustaining federally owned resources for the public's welfare is achieved.

The contractual relationship instituted by the Taylor Grazing Act's grazing leasing process was the catalyst for the well-documented historical capture of the land-use subgovernment of the BLM by organized ranching groups (Cawley, 1993; Clarke & McCool, 1996; Culhane, 1981; Davis, 1997a; Davis, 1997b; Donahue, 1999; Foss, 1960; Klyza, 1996; Knight, Gilgert, & Marston, 2002; Merrill, 2002; Nie, 2008; Smith & Freemuth, 2007; Starrs, 1998; Wilkinson, 1992). What is less well-documented is the manner in which the federal government similarly institutionalized energy minerals' (hereafter referred to as fluid minerals) leasing process, through the Mineral Lands Leasing Act of 1920, which helped establish the strong relationship between ranching interest groups and energy development interest groups. Further, what has been even less documented is how the federal government's policy of splitting the property estates of homestead properties, through the Livestock Raising Homestead Act of 1916, helped develop the muddled and tenuous legal, contractual, and regulatory relationships between the federal government, state governments of the West, ranching interests, and energy development interests.

In essence, the splitting of the property estates and their attendant property rights, combined with the regulatory leasing and permitting process and the delegation of some fluid mineral development management to the states, is a toxic stew of federal land-use policies. The toxicity of this odd blending of property rights and federal land-use policies eventually emerged as a trigger to a modern political controversy that ultimately led to the annexation of the BLM's land-use decision-making subgovernment from ranching to energy development.

Split-Estates: Interests, Common Laws, and Contracts

So what is a split-estate and why is this peculiarity of American property law and federal land-use policy of such tremendous concern to government policymakers, ranching interests, and energy development interests? First, ranchers organized early and exerted their collective influence on government land-use policies often. As mentioned above, the record of academic research clearly articulates the historical record of organized ranching groups' early development and influence. These same accounts convey the accepted conclusion of ranching's domination of decision-making subgovernments within governmental land-use agencies such as the BLM. These academic investigations only provide a nodding acknowledgement of the almost parallel historical development of the legal institutionalization of energy development. In the literature regarding federal land use, what has been written is in general agreement that organized energy development interest groups came at a much later time in American history and political development. And, unlike ranching's early efforts of organizing prior to any passage of homesteading legislation, energy development interests are organized as a response to the passage of the Mineral Lands Leasing Act of 1920 (Bradley, 1996; Committee on Onshore Oil and Gas Leasing, National Research Council, 1989; Durant, 1992; Engler, 1961; Engler, 1977; Fairfax & Yale, 1987; Flynn & Watson, 2006; Isser, 1996; Mayer & Riley, 1985; McDonald, 1979). Consequently, ranching interests firmly established themselves in the driver's seat of federal land-use policy subgovernments prior to the collective organization of energy development interests and their capacity to influence governmental land-use agencies. Simply stated, due to their early organization and capacity to control the BLM, energy development interests have historically deferred to ranching's domination of the land-use subgovernment within the BLM.

Second, common law doctrines favor ownership, management, and development of public lands and resources. For instance, since the late nineteenth century, the US Supreme Court has recognized the common law doctrine that land owners may divide their land horizontally into a surface and a subsurface estate for the purpose of economic development.[5] Once this horizontal separation occurs, legal title to the estates is vested in different owners. Courts commonly refer to the subsurface estate simply as the "mineral estate"[6] and recognize a number of implied property rights relating to the access and use of the mineral estate as belonging to the mineral estate owner (*Harris v. Currie*, 1943, as cited in Alspach, 2002).

Mineral estate owners, for example, have the implied property right to utilize "so much surface as may be reasonably necessary for operation" or "may use as much of the surface estate as reasonably necessary to produce the subsurface minerals." Additionally, mineral estate owners "have the right to enter, occupy, and make reasonable use of the surface in order to produce subsurface minerals." In terms of federal energy development, it is important to note that leases convey access and user property rights to the lessee. As one court has noted, "the reservation of the mineral estate carries the right to use as much surface as necessary to enforce the conveyed mineral estate, because if not able to use the surface lands to access the minerals below, the mineral estate would be worthless" (*Union Producing Co. v. Pittman*, 1962; *Placid Oil Co. v. Lee*, 1951; & *Harris v. Currie*, 1942 as cited by Merrill, 2002).

Exemplified by these rulings, neither federal nor state courts have typically sought to balance the rights and interests of mineral estate owners and their lessees against the rights and interests of surface estate owners, as long as the mineral estate owners' conduct in the exploration and development of the mineral estate was reasonable (King et al., 1992 as cited in Merrill, 2002; Polston, 1987). It must also be noted that conduct by energy development entities has commonly been interpreted by the courts as being reasonable if the conduct conforms to the generally accepted practices of the energy development industry (*Hunt Oil v. Kerbaugh*, 1976 as cited in Merrill, 2002).[7]

Throughout the history of energy resource development, relatively few limits have been placed on the exploration, drilling, and production operations occurring on public or private surface estate lands (*Hunt Oil v. Kerbaugh*, 1976 as cited in Merrill, 2002). Typically, most mineral estate contractual agreements or federal leases are absent components that clearly recognize the rights and interests of the surface estate owner. Federal mineral estate lessees, for example, have typically not been liable for damages to the surface estate despite the fact that subsurface mineral extraction can be incredibly damaging to the surface estate (*Hunt Oil v. Kerbaugh*, 1976 as cited in Merrill, 2002). In fact, federal and private mineral estate energy developers have historically been protected by one or a combination of the common

---

[5] Note: The *Del Monte Mining & Milling Company v. Last Chance Mining Company*, 1898, comments that "[u]nquestionably at common law the owner of the soil might convey his interest in mineral beneath the surface without relinquishing his title to the surface."

[6] Note: Courts commonly define that a horizontal severance of land creates two separate estates, including the mineral estate.

[7] Note: Common legal definition of reasonable is determined as "the reasonableness of the use of the mineral estate may be measured by the usual, customary, and reasonable practices in the industry under like circumstances of time, place, and surface estate use."

law's liberty of contracts, reasonable access, and right of first-capture doctrines. The courts' use of these doctrines, in addition to precedent rulings establishing the legal privilege of the mineral estate's dominance over the surface estate, has shielded energy developers from liability unless there is clear abuse of their right of reasonable access or their extractive operations have been conducted in a negligently harmful manner (*Hunt Oil v. Kerbaugh*, 1976 as cited in Merrill, 2002).

Historically, both federal and state administrative agencies and courts have treated the mineral estate as dominant over the surface estate, imposing few restrictions on the mineral estate (King et al., 1992 as cited in Alspach, 2002; Polston, 1987). The mineral estate owner's right to use the surface and available resources to develop the mineral estate has been termed as "right of access" and is treated by governmental agencies and courts as an easement (*Bergen Ditch & Reservoir Co. v. Barnes* as cited in Alspach, 2002, p. 91; King et al., 1992, pp. 9–2). Additionally, from the perspectives of public policy, because the legal evolution of mineral estate dominance is closely tied to economics, throughout history, development of subsurface minerals—hard rock or fluid energy—has been viewed as "essential to [the state's] comfort and prosperity," and restricting the development of the subsurface resource has been seen as "a great public wrong" (*Chartiers Block Coal Co. v. Mellon*, 1893; Berry, 1985). In short, subsurface mineral development would not be feasible without certain inherent rights of property and dominance found in the common law (Berry, 1985, as cited in Evans, 1996).[8]

At the peak of the late nineteenth-century mineral exploration and early twentieth-century industrial energy development, surface distress was typically limited to damage inflicted by picks and shovels and other items of limited technological development such as mule-drawn scrapers (*Martin v. Kentucky Oak Mining Co*, 1968, as cited in Wenzel, 1993).[9] In other words, surface damages during this period of time were generally accepted as more than a "trifling inconvenience" (Wenzel, 1993, p. 624). In fact, during this same era, the Supreme Court went so far as to essentially sanction development of the mineral estate in the center of towns if the desired mineral was located within town boundaries (*Steel v. St. Louis Smelting & Refining Co.*, 1882).[10]

Historically, and in terms of present-day court rulings, the legal issues surrounding split-estates have presented interesting dilemmas for courts and lawmakers. Specifically, because large-scale, subsurface hard-rock mineral extraction efforts did not begin in earnest until the latter part of the twentieth century, and subsurface fluid mineral energy development of energy resources in the early part of the

---

[8] Note: When Texas landowners strike water, there is surface estate remediation and legislatively enhanced liability in the oil patch. The proposal is intended for optimum protection of groundwater resources from oil and gas exploration and production in Texas.

[9] Note: The court described "usual, known, and accepted" methods around 1905 as including "picks, shovels, and slip-scrapers drawn by mules to remove the thin overburden." Overruled on other grounds by *Akers v. Baldwin*, 1987.

[10] Note: The court surmised that "to such mining claims, though within the limits of what may be termed the site of the settlement of new town, the miner acquires as good a rights as though his discovery was in a wilderness."

twenty-first century, English common law, adopted via federal legislation and state constitutions upon their admission to the Union, is ill equipped to deal with disputes arising under split-estates. As one legal scholar notes, "[t]he simple reason for [the lack of oil and gas-related legal precedent when oil and gas extraction efforts began in the United States] is that there had been no significant oil and gas development anywhere in the world prior to the latter part of the nineteenth century" (Martin, 1997, p. 312). This lack of legal guidance is in keeping with the historical accounts of how oil development interests were spurred to formal organization by the enactment of legislation directly affecting their interests. Describing the dearth of legal precedent concerning issues related to fluid mineral energy development in their 1926 treatise, Lawrence Mills and J.C. Willingham note:

> On account of its recent development, [oil and gas law] has not undergone the smelting process of the common law, which has refined and purified those branches of the substantive law that have received the consideration of Bench and Bar through the centuries. As a result, its elements are not found in the mine of adjudicated cases in a state of purity, but combined and fused with many alien principles. It is the product of case-law at its worst. (Mills & Willingham, 1926)

Therefore, federal and state courts and lawmakers have been forced over time to sort out the legal differences and emerging conflicts between competing estate development interests with virtually little guidance from precedent rulings interpreting the application of common law.

Because of, or in spite of, the lack of legal clarity, it is the economic benefit to the developer and the state that remains a central tenet in the argument for retaining legal dominance of the mineral estate. For example, a North Dakota Supreme Court described the contemporary economic-based arguments in favor of retaining mineral estate dominance over the surface estate:

> The mineral estate is dominant in that the law implies, where it is not granted, a legitimate area within which mineral ownership of necessity carries with it inherent surface rights to find and develop the minerals, which rights must and do involve the surface estate. Without such rights the mineral estate would meaningless and worthless. Thus, the surface estate is servient in the sense it is charged with the servitude for those essential rights of the mineral estate. (*Hunt Oil Co. v. Kerbaugh*, 1979)

Despite substantial developments in the technology of mineral extraction, which cause significantly greater distress to surface lands than picks and shovels, courts continue to afford mineral estate owners and developers tremendous latitude in their efforts to develop the resources within the estate. As one researcher has noted:

> [M]ineral developers now have powerful earthmoving equipment that allows a single miner to extract tons of ore per hour; new mining techniques inject cyanide and other chemicals into the ground and allow old mines to be productively reopened; increased pressures for energy sources impel production of previously 'worthless' minerals such as oil shale, lignite, methane, and geothermal steam; and new materials such as uranium have been discovered. (Wenzel, 1993, p 624)[11]

---

[11] Note: Wenzel argues that "[a]ll of these changes impinge on surface rights to an extent unfathomable in the nineteenth century."

Surface owners—typically ranchers, farmers, or homeowners—on the other hand, while faced with increasing distress upon their property and resources, are left with few remedies other than the required private contractual agreements of access and use that are negotiated between surface owner and subsurface developer. This is because modern regulation of fluid mineral development remains dependent upon the frequent application of the common law doctrine of the "liberty of contracts" in early federal legislative efforts to control the development of energy resources (Gillman, 1993). Consequently, surface estate owners fare no better at defending their interests under government regulatory remedies. As will be explored in later chapters, governmental regulation of possible remediation of expected or unexpected harms done to the surface estate and its resources remains extremely limited and ill-defined. As such, the surface owner is at the mercy of industry in their capacity to hire sound legal representation in the effort to guard against the unreasonable use of their estate by modern fluid mineral energy developers.

Finally, as the only common ground between the two parties is typically a desire for profit or the protecting of economic well-being, the contractual relationship between surface and subsurface estate owners is inherently fraught with conflict (Evans, 1996, pp. 477, 483). Additionally, as in the case of federal mineral estate ownership, leasing, and permitting, the two parties negotiating private surface access and use contracts do so under the watchful eye of the government through the tinted lenses of the BLM. As noted by one commentator, "the surface owner is often in a state of perpetual irritation at the presence of [extraction] equipment on his or her property that reduces the amount of acreage available to the owner for farming, grazing cattle, or other uses that are more desirable" (Keffer, 1994, pp. 523, 525). An additional irritant, which is viewed with great distrust by all parties engaged in the contract negotiation process, is the looming shadow of federal authority and power.

Conclusion

Documenting the historical development of ranching interests with respect to federal grazing, mining, and energy development provides the context in which a better understanding of a modern-day clash of interests occurs. Additionally, analysis of the legislative and legal manner in which these interests have, in almost parallel fashion, developed over time suggests that the clashing of interests within the land-use subgovernment of the BLM was inevitable. Understanding how federal attempts to control the activities of these interests blur with government legislation and regulatory policies establishes the conditions for the modern emergence of a political conflict, which is important. The following chapters will discuss how the struggle between the contending interests of ranching, energy development, and government will redefine the accepted academic view that the era of iron triangles and agency capture has closed.

Analysis of the historical record clearly demonstrates how the federal government's desire to manage the use of public lands during the late nineteenth and early twentieth centuries helped establish the dynamic and always evolving subgovernment relationships between the federal government, ranchers, and energy developers. And, it is in the description of this context and these dynamics that political observers can discern how these developed relationships and dormant legal conditions have come to enable the modern struggle for dominance of the BLM's land-use subgovernment.

References

Akers v. Baldwin, Ky 295, 736 SW 2d 294 (1987).

Alspach, C. M. (2002, Spring). Surface use by the mineral owner: How much accommodation is required under current oil and gas law? *Oklahoma Law Review, 55,* 89–110.

Anderson, T. L., & Hill, P. J. (1990). The race for property rights. *Journal of Law and Economics, 33*(1), 177–197.

Berry, S. J. (1985). Surface Damages in Texas: A Proposal for Legislative Intervention. *St. Mary's Law Journal., 17*(1) 121–154.

Bradley, R. L. (1996). *Oil, gas and government: The U.S. Experience* (Vol. 1). Lanham, MD: Rowman and Littlefield Publishers, Inc.

Cawley, R. M. (1993). *Federal land western anger: The sagebrush rebellion and environmental politics.* Lawrence, KS: University of Kansas Press.

Chartiers Block Coal Co. v. Mellon, *152 Pa. 286, 25 A. 597* (1893).

Clarke, J. N., & McCool, D. C. (1996). *Staking out the terrain: Power and performance among natural resource agencies* (2nd ed.). Albany, NY: State University of New York Press.

Coggins, G. C., & Wilkinson, C. F. (1987). *Federal public land and resources law* (2nd ed.). Mineola, NY: Foundation Press.

Committee on Onshore Oil and Gas Leasing, National Research Council. (1989). *Land use planning and oil and gas leasing on onshore federal lands.* Washington D.C.: National Academy Press.

Culhane, P. J. (1981). *Public lands politics: Interest group influence on the forest service and the bureau of land management.* Baltimore: Johns Hopkins University Press. for Resources for the Future.

Davis, C. (1997a). Politics and public rangeland policy. In C. Davis (Ed.), *Western public lands and environmental politics* (pp. 87–110). Boulder, CO: Westview Press.

Davis, D. H. (1997b). Energy on federal lands. In C. Davis (Ed.), *Western public lands and environmental politics* (pp. 141–168). Boulder, CO: Westview Press.

Del Monte Mining & Milling Company v. Last Chance Mining Company., 171 U.S. 55, 60 (1898).

Donahue, D. L. (1999). *The western range revisited: Removing livestock from public lands to conserve native biodiversity.* Norman, OK: University of Oklahoma Press.

Donahue, D. L. (2005). Western grazing: The capture of grass, ground, and government. *Environmental Law, 35*(4), 721–806.

Durant, R. F. (1992). *The administrative presidency revisited: Public lands, the BLM, and the Reagan revolution.* Albany, NY: State University of New York Press.

Engler, R. (1961). *The politics of oil: A study of private power and democratic direction.* Chicago: The University of Chicago Press.

Engler, R. (1977). *The brotherhood of oil: Energy policy and the public interest.* Chicago: The University of Chicago Press.

Evans, G. L. (1996). Comment: Texas landowners strike water—Surface estate remediation and legislatively enhanced liability in the oil patch—A proposal for optimum protection of groundwater resources from oil and gas exploration and production in Texas. *Southern Texas Law Review., 37*, 484–485, 515.

Fairfax, S. K., & Yale, C. E. (1987). *Federal lands: A guide to planning, management, and state revenues.* Washington D.C.: Island Press.

Flynn, A. M., & Watson, R. J. (2006). CRS report for Congress: Leasing and permitting for oil and gas development on federal public domain lands. *Congressional Research Service.* Washington D.C.: Congressional Printing Office

Foss, P. O. (1960). *Politics and grass: The administration of grazing on the public domain.* Seattle, WA: University of Washington Press.

Friedman, L. M. (1985). *History of American law* (2nd ed.). New York: Simon & Schuster.

Gates, P. W. (1968). *The history of public land law development.* Washington, D.C.: Zenger Publishing.

Gillman, H. (1993). *The constitution besieged: The rise and demise of Lochner era police powers jurisprudence.* Durham, NC: Duke University Press.

Graf, M. (1997). Application of takings law to the regulation of unpatented mining claims. *Ecology Law Quarterly, 24*, 57–60.

Hunt Oil Co. v. Kerbaugh, 283 N.W. 2d 131 (N.D. 1979)

Isser, S. (1996). *The economics and politics of the United States oil industry, 1920–1990: Profits, populism, and petroleum.* New York: Garland Publishing, Inc.

Keffer, W. R. (1994). Drilling for damages: Common law relief in oilfield pollution cases. *Southern Methodist University Law Review, 47*(523), 525.

King, C. G., et. al. (1992). Surface rights issues, state bar of Texas. *Annual Oil, Gas and Minerals Law Institute.*

Klyza, C. M. (1996). *Who controls public lands?: Mining, forestry, and grazing politics 1870–1990.* Chapel Hill, NC: The University of North Carolina Press.

Knight, C. (2002). Comment: A regulatory minefield: Can the Department of the Interior say 'no' to a hardrock mine? *University of Colorado Law Review, 73*(2), 619–626.

Knight, R. L., Gilgert, W. C., & Marston, E. (Eds.). (2002). *Ranching west of the 100th meridian: Culture, ecology, and economics.* Washington, DC: Island Press.

Large, D. W. (1986). Defining 'valuable' mineral deposits-A continuing quagmire. *Arizona State Law Journal, 1986*(3), 453–486.

Martin, P. H. (1997). Unbundling the executive right: A guide to interpretation of the power to lease and develop oil and gas interests. *Natural Resources Journal., 37*(2)311–312.

Mayer, C. J., & Riley, G. A. (1985). *Public domain, private dominion: A history of public mineral policy in America.* San Francisco: Sierra Club Books.

McDonald, S. L. (1979). *The leasing of federal lands for fossil fuels production.* Baltimore: The Johns Hopkins University Press.

Merrill, K. R. (2002). *Public lands and political meaning: Ranchers, the government, and the property between them.* Berkeley, CA: University of California Press.

Mills, L., & Willingham, J. C. (1926). *The law of oil and gas* (2nd ed.). Chicago: Callaghan and company.

Mineral lands leasing act of 1920 as amended (Title 30 of the *United States Code* §185).

Mining Act of 1866 (repealed 1872).

Nie, M. (2008). *The governance of western public lands.* Lawrence, KS: University Press of Kansas.

Polston, R. W. (1987). Surface rights of mineral owners—What happens when judges make law and nobody listens? *North Dakota Law Review., 63*, 41–42.

Smith, Z. A., & Freemuth, J. C. (Eds.). (2007). *Environmental politics and policy in the west* (Revised ed.). Boulder, CO: University Press of Colorado.

Starrs, P. F. (1998). *Let the cowboy ride: Cattle ranching in the American west.* Baltimore: The Johns Hopkins University Press.

Steel v. St. Louis Smelting & Refining Co., 106 U.S. 447, 449 (1882).

Wenzel, M. A. (1993). The model surface use and mineral development accommodation act: Easy easements for mining interests. *American University Law Review., 24*(607), 623–624.

Wilkinson, C. F. (1992). *Crossing the next meridian: Land, water, and the future of the West.* Washington, DC: Island Press.

Chapter 3
Executive Branch

Abstract President George W. Bush-Era executive actions altered federal domestic energy policies and, in turn, affected the Bureau of Land Management's (BLM) domestic energy policies and resource allocation. Executive branch actions directing the BLM to favor the energy development industry illustrate how a president and his executive appointees established, possibly unintentionally, conditions for political conflict. The unilateral use of this executive authority resulted in increased levels, numbers and types of federal energy development projects in New Mexico, Colorado, and Wyoming. And while ranchers and energy developers have a long and storied history of being closely allied in promoting and defending each other's use of federal resources for economic gain; despite this history of joint benefits, in 2001 the political-will of the Bush Administration, clearly articulated in the early use of presidential powers emphasized the development of one commodity: energy.

Keywords Executive power · Ranching · Energy · Natural resource development · Split-estate energy development

> "Bureaucrats fear that political control may facilitate agency capture, and legislators fear that agency independence may result in a bureaucracy out of control."(Spence, 1997)

This chapter presents an analysis of how modern executive branch actions have altered federal domestic energy policies and have affected the Bureau of Land Management's (BLM) domestic energy policies and resource allocation. The chapter includes an analysis of archival and government documents describing executive branch actions directing the BLM to favor the energy development industry. These events are presented chronologically to illustrate how a president and his executive appointees established, possibly unintentionally, the conditions for an impending political conflict. This chapter also documents changes to federal energy policies at the agency level that led to the reallocation of resources from ranching toward energy development. These changes resulted in increased levels and numbers and types of federal development projects in New Mexico, Colorado, and Wyoming.

© Springer Nature Switzerland AG 2019

R. E. Forbis Jr., *Altered Policy Landscapes*,

https://doi.org/10.1007/978-3-030-04774-0_3

This chapter concludes with a brief analysis suggesting that increased split-estate energy development triggered conflict and competition between the formerly allied interests of ranching and energy development.

The Willpower to Achieve a Political Objective

Prior to being sworn, President-elect George W. Bush announced his first nomination to oversee the federal energy-related administrative agencies. In December of 1999, President Bush nominated Gale Norton as the administration's new Secretary of the Interior. He did so because, mindful of his campaign promise to reduce America's dependency on foreign energy resources, President Bush and Vice President Cheney required like-minded political allies who were supportive of their strategy for expanding domestic energy development as a means of achieving national security. Thus, President Bush appointed a cadre of political and policy loyalists to leadership positions within the federal administrative units charged with the management and oversight of domestic energy development.

These political loyalists were expected to take an executive-led energy task force's recommendations, apply the president's executive orders, and increase the nation's supply of domestic energy resources. If successful, the administration would fulfill the promise of reducing the country's dependency on foreign energy resources, achieve a measure of national security, and secure a political victory for the administration. This is because political appointments, like task forces and executive orders, are expressions of a presidential willpower in their ability to wield direct influence on existing legislation and administrative processes. As it turned out, the Bush Administration was exceptionally adept at expanding and using executive power. And, over the course of Bush Administration's two terms in office, the administration's political appointees would faithfully carry through with the implementation of the Bush-Cheney domestic energy plan.

With any alteration to federal land-use management practices, there is always the potential to destabilize working relationships between stakeholders. This is particularly true of federal land management agencies, ranchers, and energy developers. What is also true is that ranchers and energy developers have a long and storied history of being closely allied in promoting and defending each other's use of federal resources for economic gain. Indeed, the capacity to bring cattle, minerals, and energy resources to the market as public commodities is considered a legitimate and beneficial use of public lands and resources. Despite this history of joint benefits, in 2001, the political will of the Bush Administration, clearly articulated in the early use of presidential powers, emphasized the development of one commodity: energy.

Asserting Political Control over Administrative Decision-Making

Known simply as the "delegation problem," the efficacy of political control over bureaucratic decision-making remains a debated topic among political scientists (Moe, 1993; Moe & Howell, 1999; Spence, 1997). While the majority of scholarship explores congressional efforts to control the bureaucracy (Moe, 1993, Moe & Howell, 1999, Spence, 1997), there is also a comprehensive body of literature devoted to exploring presidential efforts of political control (Moe, 1993, Moe & Howell, 1999, Spence, 1997). However, no matter the institutional location of politically motivated efforts to achieve bureaucratic control, attempts to measure the resulting efficacy with any degree of accuracy have met with mixed results.

As David B. Spence argues, neither theoretical positivists nor quantitative empiricists have "demonstrated that politicians can overcome the delegation problem" as they have a tendency to "model the problem away" in one of two ways (Spence, 1997, p. 199). According to Spence, researchers have a tendency to presume that political control is exerted in either ex post or ex ante fashion. On the one hand, positive theorists tend to overemphasize political control as a matter of ex post political oversight of bureaucratic agencies' procedures and processes. On the other hand, quantitative empiricists overemphasize the dependent variable of ex ante political control as a matter of the bureaucratic agencies' enabling legislation. In either case, Spence argues, because researchers "overestimate the degree to which political control occurs," they can offer neither explanation nor prescription to the "delegation problem" (Spence, 1997, p. 215). The problem, according to Spence, is that in measuring the impact of political control, researchers have missed the critical distinction between "policy making" and "policy implementation" (Spence, 1997, p. 212).

Spence concludes his argument with the assertion that "If the technologies of social scientific investigation have trouble accounting for the complexity of agency policy choice, we must improve existing technologies or find new ones" (Spence, 1997, p. 215. While Spence's argument is justly critical of this body of work, his recommendation for improvement relies on the hope of technological advancements in the singular methodological realm of quantitative-based research. This rather narrow approach ignores the promise and possibility of improving this area of scholarship via other methodological means, including qualitative-based methodological research.

Methodological choices aside, qualitative researchers have not fared any better in their attempts to describe fully and account for the effectiveness of elected officials to politically control the bureaucracy. This is not to say that there have not been important and significant qualitative contributions made in the institutional study of the relationship between Congress and the bureaucracy or, for that matter, the presidency and the bureaucracy. Researchers have made significant strides in providing rich descriptive narratives of these complex institutional relationships. Spence's critique of quantitative research in this field of inquiry is equally applicable to similarly oriented inquiries of a qualitative nature. This is because they too suffer from the same malaise articulated by Spence in his criticism of theoretical positivists' and quantitative empiricists' inquiries.

The "delegation problem" debate and the issues associated with it will not be resolved here. What follows instead is an attempt to provide a descriptive narrative tracing the causal pathway through which the Bush Administration utilized the unilateral nature of presidential power in order to politically control the BLM and alter its energy policies. In essence, changes in the executive branch led to changes in domestic energy policy. The changes described here do not underestimate the effect of technological breakthroughs in domestic energy exploration and development, nor do they underestimate the impact of economic conditions affecting the price of energy. These effects will be discussed in later chapters. This chapter is, in a manner of speaking, a narrative measure of how politically effective the Bush Administration was in its strategic use of executive powers to politically control the BLM in the attempt to successfully achieve its desired political objectives.

Executive Power and the Capacity to Affect Administrative Change

In his analysis of the Nixon and Reagan administrations, Richard P. Nathan argues that "elected chief executives--presidents, governors, mayors--and their appointees should play a larger role in administrative processes" (Nathan, 1983, p. vii). The argument's premise, "management tasks *can and should* be performed by partisans," hinges on Nathan's belief in the executive branch pursuing implementation of its policy objectives through the strategic use of executive power (Nathan, 1983, p. 7).

According to Nathan, the use of executive power is legitimate so long as the executive's policy objectives are carried through within the confines of existing legislation and administrative procedural processes. Nathan's prescription for the realization of this strategy is one of political delegation. This means that strategic delegation of executive authority is a manifestation of presidential influence. This influence can affect a bureaucratic policy domain in a manner that "penetrates the [domain's] administrative process" (Nathan, 1983, p. 82). This type of political authority is necessary because, as Nathan argues, "in a complex, technologically advanced society in which the role of government is pervasive, much of what we would define as policymaking is done through the execution of laws in the management process" (Nathan, 1983, p. 82). The message to presidents here is straightforward; in order to successfully achieve executive policy objectives, it is imperative that a president wield the tools of executive authority within the confines of existing legislation and administrative procedures in a manner that influences agency-level decision-making.

In a follow-up to Nathan's research, Robert F. Durant's (1992) account of the Reagan administration is notable for its narrow focus on a single politically oriented policy objective: altering federal resource management in a manner favoring economic development. Durant's investigation of Reagan's strategic administrative efforts to effect changes in how federal resources were managed by the DOI and the BLM in particular finds that the Reagan administration's efforts were in the end highly ineffectual. Notably, Durant's findings imply that one reason for the Reagan administration's failures is that within the BLM, there existed a deeply entrenched and resourceful subgovernment that sought to protect the agency's status quo. As Durant argues:

> [The] political us[e] of the administrative presidency to reorient policy...to alter bureaucratic agendas substantially... [is] unlikely to find an agency's 'dominant coalition' predisposed to change. Coalition members are prone to buffer organizational cores from such 'turbulence' and to protect their organization's fragile political economy. Equally unsympathetic to change are clienteles accustomed to existing agency rules, relationships, and largesse. This, in turn, makes policy initiatives distinctly vulnerable to fire alarm oversight, with the type of agenda item pursued by [political officials] conditioning the nature, scope, and intensity of resistance mounted by opponents. (Durant, 1992, p. 238)

Durant notes that any future attempts to untangle the dynamics of agency subgovernments in the face of a sustained political effort to alter the existing policy orientation of administrative agencies should embrace the "validity of the causal theory" (Durant, 1992). Thus, clearly accounting for causal pathways between executive power and administrative agencies will help clarify the means by which presidents pursue political control of bureaucratic decision-making. In turn, clarifying the causal pathways of executive influence will assist in capturing the dynamic of strategic actions among subgovernment actors as they seek to sustain the decision-making status quo.

Durant concludes by cautioning that establishing bright-line causal pathways of political control over administrative procedures in the course of implementing policy objectives is dependent on the ability to account for the inherent characteristics of "bureaupolitical dynamics during implementation" (Durant, 1992, p. 238). Here, Durant argues that two characteristics of bureaupolitical dynamics condition any success for politically controlling the implementation of executive policy objectives: (1) validity of the novel policy initiative and (2) softening of policy communities and larger publics over time (Durant, 1992, pp. 238–239). Durant argues:

> In the real world, of course, these two variables can interact to produce distinct bureaupolitical dynamics...however, the bureaupolitical politics occasioned are not 'caused' by the interaction of the two variables...rather, their interaction either affords or constrains opportunities for challenge to those opposed to drastic policy reorientation. (Durant, 1992, p. 239)

Both Nathan and Durant's research efforts illustrate the strategic use of broad executive power and its potential to affect administrative decision-making and subgovernment activity. Nevertheless, Nathan's research remains a narrative prescriptive bordering on a polemical treatise. And while Durant's research accounts for interest group efforts to maintain the status quo in the face of the Reagan administration's attempt to politically affect a shift in federal land-use management within the BLM, his effort focuses on finding the degree of effectiveness in the executive's realization of favored land management policy objectives. As such, the turbulence caused by Reagan's executive actions and their effect on the existing coalition of interest groups that constitute the BLM's land-use subgovernment is never fully articulated. Thus, as Durant himself notes, "the types of policy initiatives, bureaucratic responses, and political dynamics outlined are hardly exhaustive, must be further elaborated, and require empirical testing" (Durant, 1992, p. 321).

Political science scholarship concerning the influence of the president is wide-ranging. Beginning with Richard Neustadt's (1960) argument that presidential power is reflected in the ability to influence, other political scientists have sought to extend our understanding of presidential power and its impact. Since Neustadt, researchers have sought a better understanding of presidential power by investigating a variety of presidential initiatives to strengthen their control over administrative agencies. They have done so through a variety of means, e.g., personnel management, appointments, White House staffing, reorganization, assertion of legal prerogatives, executive orders, signing statements, etc. (Pfiffner, 1999). The ability of President George W. Bush to disrupt a relatively stable land management subgovernment presents a unique opportunity to understand the impact of presidential power.

The Election of President George W. Bush

With the election of George W. Bush in 2000, the government of the United States undertook an ambitious approach in responding to the energy needs of the nation. From the time of the presidential campaign to the election, President Bush promised the American public a policy initiative to address the nation's growing demand for energy and secure energy independence. The Bush Administration often premised its argument for securing the nation's energy resources and independence on the basis of strengthening national security. With the terrorist attack of September 11, 2001, the administration's argument gained substantial validity in the minds of elected officials, the policy community, and the general public.[1]

The Bush Administration utilized the increased level of support for its argument and strategically wielded executive power in a manner that would advantage exist-

---

[1] Note: Gallup poll of May 23, 2001a shows public support for the Bush energy plan at 44%. Public belief in the Bush energy plan's success was 65%. Gallup poll of June 5, 2001b shows public concern over energy resources as America's most important and pressing problem at an historic high of 58%. Gallup poll of April 3, 2002 shows overall public approval for President Bush's handling of energy policy at 57%. Gallup polling data (March, 2001–2003) shows public opinion that the United States will face critical energy shortages over the next 5 years, as 60% (March 2001), 48% (March 2002), and 56% (March 2003).

ing legislation and administrative processes to achieve the objective of expanding domestic energy resource development. The tragedy of September 11, 2001 was, in many ways, simply a fortuitous event allowing the Bush Administration to successfully implement the means for achieving its energy policy objectives. Thus, with the support of like-minded congressional leadership, and over the course of their 8 years in office, the Bush Administration successfully implemented a series of political and administrative strategies that resulted in (1) a shift in domestic energy policy, (2) the creation of a political conflict between powerful interest groups, and (3) the disruption of a long static subgovernment within the Bureau of Land Management.

President Bush's Energy-Related Political Appointments at DOI

President Bush's choice of Department of the Interior (DOI) nominees was a direct reflection of his administration's desire to expand domestic energy. Most significant among the president's "energy nominees" was Gale Norton to head the Department of the Interior. As the president's nominee, Secretary Norton's history of professional and political accomplishments was notable for their consistent support and defense of deregulation and free-market principles in the management of federal lands and resources. A former DOI attorney under President Reagan's controversial and short-lived Secretary of the Interior James Watt, Norton's nomination was met with great cheer from conservative free-market thinkers as well as industry representatives of the timber, mining, and energy development lobby (Jehl, 2000).

To others, most notably members of the environmental protection community, Ms. Norton's nomination was greeted with dismay. As the national spokesman of the Sierra Club, Allen Mattison, famously remarked, "Our view is that she's James Watt in a skirt" (Jehl, 2000).

Mentored by Watt during her tenure at the politically conservative Mountain States Legal Foundation (MSLF), Norton was a true believer in the pro-development management principle for public lands and resources. Other important Bush-Cheney DOI political appointees had similar backgrounds. For example, following Gale Norton's appointment, President Bush nominated another Reagan-Watt Era alumni, J. Steven Griles. As Undersecretary of the Interior, Griles was second only to Norton in the chain of political authority being assembled at DOI. Under Secretary Watt's tenure at DOI and afterward, Griles served as Deputy Director of the Office of Surface Mining and as Assistant Secretary and Deputy Assistant Secretary of the Interior for Land and Minerals Management. It is important to note that it was Mr. Griles who, in anticipation of his Senate confirmation, served as the DOI's representative during the course of the Cheney Energy Task Force deliberations in 2001 (US Department of the Interior, 2001).

President Bush also nominated Rebecca Watson as Undersecretary of the Interior for Land and Minerals Management. Having served as assistant general counsel for energy policy at the Department of Energy (DOE) in the previous Bush Administration, Ms. Watson was a former law school classmate of Secretary Norton's and, at the time of her nomination, a MSLF colleague of both Norton and Watt. With the Senate's approval of Watson's appointment, she was charged with administrative and managerial responsibility for the Bureau of Land Management, the Minerals Management Service, and the Office of Surface Mining Reclamation and Enforcement (US Minerals Management Service, 2002).

Finally, President Bush nominated Kathleen Clarke as Director of the Bureau of Land Management. At the time of her appointment, Clarke served as Executive Director of Natural Resources for the State of Utah under then Governor Michael Levitt (Gov. Levitt would himself become President Bush's nominee as Secretary of Health and Human Services). Prior to her appointment as Utah's Director of Natural Resources, Clarke served as a member of Rep. James Hanson's (R-UT) administrative staff. Rep. Hanson, a conservative, was himself a fervent legislative advocate of developing resources on public lands and vocal champion of "sagebrush rebels" (Spangler, 2001). Senator Hanson would serve as Chair of the House Committee on Natural Resources during the early years of the Bush Administration when the expansion of domestic energy development was beginning to gain political and popular support (Neustadt, 1960).

Given the Bush Administration's broad policy objective of achieving national security by means of energy independence, these appointments were not the only political appointments with professional ties to varied energy-related development entities. Throughout the federal government, Bush-Cheney political appointees with ties to the energy industry or other extractive industries dominated energy and environment-related administrative agencies. The extent to which the administration's appointees were tied to the energy lobby was so profound that the administration is often referred to as the "oil and gas administration" (Finley, 2003).

The administrative hierarchy of federal agencies charged with managing the nation's energy, environmental, and public lands and resource-related policies from the president and vice president down was dominated by former fossil fuel energy development executives, attorneys, and lobbyists. With the history of political defeat suffered by previous administrations' failure—most notably the Reagan administration's failure (Durant, 1992; Nathan, 1983)—to expand domestic energy development, the choice for the new Bush Administration was clear: use executive power to effect a shift in the political leadership of administrative agencies, charge them with implementing executive policy directives to facilitate change in existing energy policy, and expand domestic energy resource development.

Vice President Cheney's Energy Task Force

Chaired by Vice President Cheney, meetings of the "National Energy Policy Development Group" were by invitation only and conducted behind closed doors. Aside from invited members from the newly elected administration and America's leading energy producing companies, no stakeholders participated in these strategic discussions. Indeed, these discussions were so secretive in nature that the administration resisted General Accounting Office (GAO) and nonprofit organizations' attempts to force the public release of the group's member list and meeting transcripts. And, although almost 40 task force meetings with industry representatives took place, the Bush Administration successfully resisted the official release of any information concerning task force members or the closed-door policy discussions. The administration's resistance was validated in 2005 when the US Federal Court of Appeals for the District of Columbia ruled unanimously in favor of the administration's "executive privilege" argument for not releasing any internal documentation regarding the energy task force (Abramowitz & Mufson, 2007; Judicial Watch Press Office, 2005).

Controversy notwithstanding, the Bush-Cheney policy development group issued its final report to the president and the public on May 16, 2001. The report, entitled "National Energy Policy," detailed the administration's energy plan and offered strategies for its implementation (National Energy Policy Development Group, 2001). Within 2 days of the report's release, President Bush issued two executive orders (E.O. 13211 and E.O. 13212) charging federal agencies to facilitate and expedite the means by which the expansion of developing America's domestic energy resources would be achieved. In essence, these executive orders signified that the report's findings had been implicitly accepted and strategies for its implementation had been adopted by the administration (Mayer, 2001).[2] While the executive branch's overarching objective was to increase the development of domestic energy resources, achieving that goal was a daunting task. As suggested by Nathan (1983), in order to meet the overarching objective, the administration would have to directly engage existing legislation in a manner that would affect change in the administrative processes of federal agencies to hasten the desired expansion of domestic energy resource exploration and development.

A key element to the success of the political objective was the administration's ability to move the bureaucracy and expand access to federally administered lands and resources. Moving the bureaucracy would require altering the procedural processes for leasing public lands and issuing approved permits to drill (APD). Expanding access to federal lands and resources required that the administration make a choice between two political strategies (Howell, 2005).[3] One political strategy was to simply send the "National Energy Policy" to the Congress for legislative

---

[2] Note: Mayer argues that the presidential power to control the actions of executive agencies is manifest in the issuing of executive orders. Mayer finds that executive orders are an expression of political will in the face of an intractable or indecisive Congress and that executive orders enhance bureaucratic accountability by creating a clear decision trail that leads directly to the president.

[3] Note: Howell argues that in order to "advance their policy agenda, presidents have two options. They can submit proposals to the Congress and hope that its members faithfully shepherd bills into laws; or they can exercise their unilateral powers--issuing such directives as executive orders, executive agreements, proclamations, national security directives, or memoranda--and thereby create policies that assume the weight of law without the formal endorsement of a sitting Congress," pp. 417.

deliberation and action.[4] The other was to wield executive power in a manner that would facilitate executive implementation of the energy plan. Given the legislative history of defeat suffered by energy interests to expand domestic energy development, the choice for the administration was clear: use executive power to affect a shift in federal energy policy via political appointments and then issue executive orders directing agencies charged with administering domestic energy development to alter their administrative processes.

Executive Orders 13211 and 13212

As noted earlier, the Bush Administration issued executive orders 13211 and 13212 on May 18, 2001. These executive orders directed all federal land management agencies—particularly the BLM—to expedite the leasing of federal lands for energy development and the approval of existing, and future, approved permits to drill (APD). Executive order 13212, entitled "Actions To Expedite Energy-Related Projects," directed federal agencies—particularly the BLM—to "expedite their review of permits or take other actions as necessary to accelerate the completion of such [energy-related] projects." Executive order 13212 also ordered the establishment of an interagency task force, chaired by the chairman of the Council of Environmental Quality, "to monitor and assist the agencies in their efforts to expedite their review of permits or similar actions, as necessary, to accelerate the completion of energy-related projects, increase energy production and conservation, and improve transmission of energy" (Finley, 2003).[5] Finally, executive order 13212 directed the interagency task force to "monitor and assist agencies in setting up appropriate mechanisms to coordinate Federal, State, tribal, and local permitting in geographic areas where increased permitting activity is expected" (Executive Order 13212, 2001).

Entitled "Actions Concerning Regulations That Significantly Affect Energy Supply, Distribution, or Use," executive order 13211 required that all federal agencies "prepare a Statement of Energy Effects when undertaking certain agency actions." And, as described in executive order 13211, these Statements of Energy Effects were intended to:

> …describe the effects of certain regulatory actions on energy supply, distribution, or use…
> [And] consist of a detailed statement by the agency responsible for the significant energy
> action relating to: i. any adverse effects on energy supply, distribution, or use (including a

---

[4] Note: The Bush Administration did eventually realize legislative success for their domestic energy strategies and policies. The Energy Act of 2005 was passed and signed into law by President Bush. As some have noted, the net effect of the Act was an affirmation of the administration's actions to bring about the expansion of domestic energy development.

[5] Note: Another Bush-Cheney appointment with ties to extractive industries, the Chair of the White House Council on Environmental Quality was James Connaughton, legal counsel for General Electric and Atlantic Richfield, and their challenge to the EPA's directive was regarding responsibility for cleanup of superfund sites.

shortfall in supply, price increases, and increased use of foreign supplies) should the proposal be implemented, and ii. reasonable alternatives to the action with adverse energy
effects and the expected effects of such alternatives on energy supply, distribution, and use.
(Executive Order 13211, 2001)

These two executive orders sought to comprehensively change existing federal
energy policy and administrative processes within land and resource agencies.

It has been argued that most executive-led strategic efforts to influence policy
change within administrative agencies or their decision-making subgovernments
cost too much political capital given the relatively modest levels of success of
those efforts (McCool, 1989). Still others have argued that, as executive orders go,
most presidential policy directives are relatively unnoticed as the change they
affect is limited to the administrative agency targeted by the president (Durant,
1992; Mayer, 2001). In the case of executive orders 13211 and 13212, there was
not much political capital to spend as the president was just months from being
sworn in and the administration signaled the opening move in its effort to control
the BLM's energy policies and administrative procedures. Quite simply, executive
orders 13211 and 13212 should be considered one piece among the many political
strategies employed as a means of achieving administration's domestic energy
policy objectives. These executive orders are notable because within the newly
released National Energy Policy, 105 recommendations had been designed specifically to increase domestic energy development, and among those recommendations, 73 could be implemented via presidential directives to energy-related
agencies, while the remaining 32 required the Congress to pass new legislation or
amend existing laws (Longley, 2001).

The BLM Responds to Change in the Executive Branch

The bureaucratic response to the unilateral use of executive powers was immediate. Within roughly 2 years of President Bush's political appointments being in
office and his issuing of executive orders 13211 and 13212, the BLM began the
process of changing its existing energy policies to reflect the political goal of
expanding domestic energy development. On August 8, 2003, BLM Director
Kathleen Clarke notified state and field offices that implementation of President
Bush's National Energy Policy would begin immediately. The new administrative
management policies instructed all BLM offices and land-use planners to reduce
or eliminate regulatory impediments to oil and gas leasing and production on
BLM lands. The Director's order instructed BLM staff to concentrate their efforts
on what Clarke had designated as "focus areas" where the potential for oil and
gas development was high. The order also instructed BLM field managers to pri-
oritize work-related efforts that would promote oil and gas planning, leasing, and
permitting (Longley, 2003).

In issuing the directive Director Clarke established the deadline of December 31,
2003 for BLM personnel to evaluate and report the need to change "existing land-

use plans to facilitate oil and gas exploration and development" in accordance with Energy Policy and Conservation Act of 2000 (Longley, 2003). In establishing new energy policies, BLM land-use planners were instructed to act in a manner that would "not unduly restrict access to federal lands, while continuing to protect resources when they review[ed] oil and gas lease stipulations, especially in those cases where an unnecessary stipulation could result in the abandonment or delay of a project" (Longley, 2003). Finally, Director Clarke's order required all BLM state offices with significant energy-related programs "to conduct at least one meeting with industry representatives" within a year of the directive's issuance to "share findings and discuss oil and gas related policy changes" (Longley, 2003).

Clearly, a change in the presidency led to a change in domestic energy policy from within the DOI and more importantly, throughout the BLM. As most field offices with significant oil and gas development projects are located throughout the Rocky Mountain West, the directive had its greatest effect in the states of New Mexico, Colorado, Wyoming, Montana, and Utah. As a result, the easing of oil and gas development regulations and administrative oversight, as well as prioritizing oil and gas activity, in 2003, triggered a modern energy boom throughout the states of the American West. This was particularly true of energy resource development in the form of coalbed methane (CBM) natural gas.

One example of how quickly the administration was realizing success in achieving its policy objective is taken from the Wyoming State Office of the BLM and the Wyoming Oil and Gas Conservation Commission. In 2003, 39,000 CBM approved permits to drill (APDs) were issued by the State of Wyoming. These 39,000 APDs represented an average of 18 permits being approved per day and an average of 7 wells being drilled per day throughout the State of Wyoming. Additionally, the rate of permit hearings in Wyoming increased that year as well. In 2003, the state's oil and gas commission held 814 area drilling permit hearings with 55% of those hearings concerning the exploration and development of CBM. These hearings resulting in 900 individual parcel drilling permits being issued by the State of Wyoming. Respectively, the state's 814 area development permit hearings represented a 100% increase over the previous 5 years with—what was at the time—an expectation that permit hearings would again experience a 100% increase in 2004. Additionally, the 900 individual parcel drilling permits issued in 2003 represented a 100% increase from the previous 30 years, and they too were expected to experience a 100% increase in 2004 (Likwartz & Parfitt, 2004).

The Wyoming Oil and Gas Conservation Commission and the Wyoming BLM estimated in 2004 that the agency would, until the year 2014, issue an additional 71,000–76,000 drilling permits for the exploration and development of oil and natural gas in the State of Wyoming. At the time, those estimates stood in stark contrast to the documented 70,000 drilling permits the State of Wyoming had issued since its statehood in 1890 (State of Wyoming, n.d.; Bureau of Land Management, n.d.). While these numbers provide evidence only of the State of Wyoming's unprecedented level of oil and gas exploration and development, the numbers were indicative of what was occurring throughout the states of the Rocky Mountain West.

Expanded exploration and drilling were not the only energy-related activities affected by the BLM's change in energy policies. For example, the BLM also expanded its energy leasing activities in accordance with the new administrative directives and policies. One example of the early nature of expanding energy leasing is the Utah State Office of the BLM's June of 2004 energy lease auction for the exploration and development of the subsurface energy resources across 281,000 acres in the State of Utah. As was reported by *The Salt Lake Tribune*, "the federal government set a record with its June oil and gas auction in Utah…as part of the Bush Administration's push toward domestic energy production…Records were made to be broken, though… The next quarterly lease auction slated for September 8 easily outpaces the June sale, with 362,665 acres spread across 223 parcels" (Nailen, 2004).

All across the West, record numbers of APDs were being issued by state energy commissions and the BLM, and a record number of acres were being offered by the BLM. The observation among those most directly affected by the change in federal energy policy was this: a change in the executive had led, successfully, to a change in domestic energy policy.[6] These executive-led changes to the BLM's traditional energy policies had the unintended consequence of establishing the conditions required to trigger a political conflict between the industries of ranching and energy.

Policy Change Triggers Political Conflict

The exploration and development of energy resources occurs primarily on federal public lands. As one would expect, most fluid mineral extraction has and continues to take place on public lands administered by the BLM. It has been relatively well-documented that the use of public lands for the extraction of energy resources is but one of the traditional uses of public lands. And, it is also relatively well-documented that grazing is also one of, if not *the paramount* traditional use of, public lands. As such, the large industry users of public resources—ranching and energy—have a long and storied history and tradition of cooperation in the West. The practice of both grazing and energy development on public lands—while controversial to some—has not, generally speaking resulted in political conflict. Both industries, at their core, shared the belief that the public lands were managed in such a manner that results in the greatest economic benefit to the user. Access to public lands, like the use of the public domain, was an issue for collaborative decision-making between the interests. And, when it came to energy development, it was an unstated agreement among the interested parties that the interests of the rancher were respected. Essentially, the political history and traditions of the West made it clear that when it came to energy and cattle, cattle came first. Not unlike the manner in which ranchers view Western water rights, ranchers believed that the principle of "first in use, first in right" also applied to Western public lands.

Despite this history, the Bush Administration's objective of expanding domestic energy development activity presented a challenge to the traditions and beliefs. Whether that challenge was intentional or not may never be fully known. But what

---

[6] Note: See generally discussion in Chaps. 5, 6, and 7.

is clear today is that by 2003, energy development activities had begun to substantially interfere with, and disrupt, ranching activities. The subsurface development of domestic energy resources had begun not only to encroach upon the traditional grazing areas of the public domain, but it had also begun to encroach upon the privately owned ranches of the West.[7]

In the Western United States, energy exploration and development can occur not only on public lands but also on private lands. The practice of developing the so-called split-estate for energy resources has become commonplace.[8] In turn, this practice increasingly placed the interests of ranchers in conflict with the energy industry. As industry more frequently sought access to develop federal energy leases located on privately owned ranch lands, ranchers began to seek remedies that would prohibit industry's access and development activities on their property. In turn, ranchers and ranching organizations sought relief from the BLM as the agency was responsible for managing the energy leases as well as regulatory oversight of energy-related activities.

The aggressive nature of the Bush Administration's domestic energy policies had awakened the dormant, but inherently conflict-ridden federal legislation of homesteading, mining, oil and gas, and grazing. The politically motivated expansion of modern domestic energy resource development had the effect of creating conflict between the principal actors—ranchers and energy—within the land-use subgovernment of the BLM. Nothing less than control over the direction of the federal government's land-use policy decisions was at stake. The stakes in the outcome of the conflict were enormous for both interest groups, and depending on which side won, it was expected that the winner would emerge as the dominate force over all other uses or future uses of the public domain. Simply put, the conflict's outcome held the potential to shift the operating paradigm of the BLM's decision-making subgovernment and, in turn, America's public lands and resources policy.

Conclusion

President George W. Bush and Vice President Cheney, former executive officers of energy development companies from energy-producing states of Texas and Wyoming, respectively, embarked upon an executive-oriented strategy to increase domestic energy production from the time of their election to office. Prior to their inauguration, and in the early years of their administration, they justified this

---

[7] Note: See generally discussion in Chaps. 5, 6, and 7.

[8] Note: The practice of developing the split-estate energy resources is estimated to be 3–5% (1,740,000–2,900,000 total acres) of all energy activity within the intermountain states that compose the Rocky Mountain West. 58 million acres across five Western states: New Mexico, Colorado, Wyoming, Montana, and Utah. Each state has roughly 10–12 million split-estate acres. On average, each state would have 300,000–500,000 split-estate acres in development. The BLM does not keep specific data regarding split-estate energy development. See generally Bureau of Land Management, 2007.

strategy by arguing that without a significant increase in domestic energy development, the nation's national economy and security were at risk. And, fortuitously, a series of events occurred during the course of their first term that effectively solidified the administration's argument in the minds of the American public.

Spurred by global conflict and the growing economic power of global rivals helped establish conditions for unprecedented increases in the market price for global energy resources. Combined with the tumultuous global events of the day, the steady increase in the price of energy resources, particularly the market rate for a barrel of oil, profoundly affected the American psyche. Indeed, America's military engagement in widespread global conflict, as well as contending with emerging foreign economic powers and their competition for energy resources, spurred an almost daily rise in the price of energy resources.

As these events unfolded, they were daily fodder for all the major American news outlets. In turn, the American public responded and viewed the administration's efforts to expand domestic energy production as necessary for securing the nation's economic and national security interests. In essence, those who would engage in activism to slow the Bush Administration's efforts to expand domestic energy development were effectively marginalized.

With the benefit of hindsight, the public is now coming to reflect upon the consequences resulting from the administration's sustained efforts to expand domestic energy production. As has been suggested, one consequence of these efforts was the awakening of a long-dormant legislative history that would trigger an unexpected political conflict. This political conflict, addressed in the next chapter, severely strained the traditional alliance between ranching and energy development interests and effectively altered the domination of the BLM's land-use decision-making subgovernment.

References

Abramowitz, M., & Mufson, S. (2007, July 18). Papers detail industry's role in Cheney's energy report. *The Washington Post*. Retrieved from http://www.washingtonpost.com/wp-dyn/content/article/2007/07/17/AR2007071701987.html?hpid=topnews

Bureau of Land Management. (2007). *Surface operating standards and guidelines for oil and gas exploration and development*. Retrieved from http://www.blm.gov/wo/st/en/prog/energy/oil_and_gas/best_management_practices/gold_book.html

Bureau of Land Management. (n.d.). Wyoming: In the spotlight. Retrieved from http://www.wy.blm.gov

Durant, R. F. (1992). *The administrative presidency revisited: Public lands, the BLM, and the Reagan revolution*. Albany, NY: State University of New York Press.

Executive Order No. 13211, 66 C.F.R 28355 (2001).

Executive Order No. 13212, 66 C.F.R. 28357 (2001).

Finley, D. S. (2003). The Bush appointees. Retrieved from http://restoringsanity.org/politics/bush_appointments.html

Gallup News Service Poll, (2001a, May 23). Public divided in reaction to Bush energy plan: Many feel it does not do enough to conserve or produce energy. Retrieved from http://gallup.com/poll/4438/Public-Divided-Reaction-Energy-Plan.aspx

Gallup News Service Poll, (2001b, June 5). Energy crisis 2001: Where America stands: 10 findings about American public opinion on energy. Retrieved from http://www.gallup.com/poll/4441/Energy-Crisis-2001-Where-America- Stands.aspx

Gallup News Service Poll, (2003a, April 3). Public rates Bush highly on terrorism, defense issues: Gets lowest approval ratings on Social Security, campaign finance. Retrieved from http://www.gallup.com/poll/5563/Public-Rates-Bush-Highly-Terrorism-Defense-Issues.aspx

Gallup News Service Poll, (2003b, March 13). Americans foresee energy shortage within 5 years: Majority favors conservation over production as energy solution. Retrieved from http://www.gallup.com/poll/7987/Americans-Foresee-Energy-Shortage-Within-Years.aspx

Howell, W. G. (2005). Unilateral powers: A brief overview. *Presidential Studies Quarterly., 35*(3), 417–439.

Jehl, D. (2000, December 30) The 43rd President; Interior choice sends a signal on land policy. *The NewYork Times.* Retrieved from http://www.nytimes.com/2000/12/30/us/the-43rd-president-interior-choice-sends-a-signal-on-land-policy.html

Judicial Watch. (2005, May 10). Appeals court permits energy task force records to remain secret. Press release. Retrieved from http://www.judicialwatch.org/5442.shtml

Likwartz, D., & Parfitt, T. (2004). *Split-Estates*. 2004 BLM National Fluid Minerals Conference, June 22, 2004, Cheyenne, Wyoming.

Longley, R. (2001, May 17). Bush energy policy: Make more, use less. Retrieved from http://usgovinfo.about.com/library/weekly/aa051701a.htm

Longley, R. (2003, August 8). Oil and gas development regulations eased. Retrieved from http://usgovinfo.about.com/b/2003/08/08/oil-and-gas-development-regulations-eased.htm

Mayer, K. R. (2001). *With the stroke of a pen: Executive orders and presidential power*. Princeton, NJ: Princeton University Press.

McCool, D. C. (1989). Subgovernments and the impact of policy fragmentation and accommodation. *Policy Studies Review, 8*(4), 264–287. https://doi.org/10.1111/j.1541-1338.1988.tb01101.x

Moe, T. M. (1993). Unilateral action and presidential power. *Presidential Studies Quarterly, 29*(4), 850–872.

Moe, T. M., & Howell, W. G. (1999). The presidential power of unilateral action. *The Journal of Law, Economics, and Organization, 15*(1), 132–179.

Nailen, D. (2004, August 31). Oil, gas auction breaks records. *Salt Lake Tribune*: Article ID: 105278D64C96A0DA.

Nathan, R. P. (1983). *The administrative presidency*. New York: Wiley.

National Energy Policy Development Group. (2001). National energy policy. Washington D.C.: U.S. Government Printing Office. Retrieved from http://www.whitehouse.gov/energy/National-Energy-Policy.pdf

Neustadt, R. E. (1960). *Presidential power: The politics of leadership*. New York: Wiley.

Pfiffner, J. P. (1999). *The managerial presidency*. College Station, TX: Texas A&M University Press.

Spangler, D. K. (2001, December 22). Utahn ok'd to lead BLM: Clarke is first woman in post, 2nd from state. *Deseret News*. Retrieved from www.deseretnews.com/article/881107/

Spence, D. B. (1997). Agency policy making and political control: Modeling away the delegation problem. *The Journal of Public Administration, Research, & Theory, 7*(2), 199–219.

State of Wyoming. (n.d.). Wyoming oil and gas conservation commission (web page). Retrieved from http://wogcc.state.wy.us

United States Department of Interior. (2001). Biographical information: Steven Griles. Retrieved from http://www.doi.gov/bio/griles.html

United States Minerals Management Service. (2002). Rebecca W. Watson. Retrieved from www.mms.gov/mmab/Archives/policy-committee-archives/Meetings/Spring02/BioRebeccaWatson.PDF

Chapter 4
Subgovernments

Abstract Changes in federal domestic energy policy resulted in increased split-estate energy development. Documented from primary government documents and journalistic sources across three settings, New Mexico, Colorado, and Wyoming, where the vast majority of split-estate energy development and conflict between ranching and energy development interests occurred, analysis of the data shows that the expansion of split-estate energy development raised the level of frustration with the BLM among ranchers. Their frustration is documented as a spiraling conflict via accounts of ranchers petitioning their state legislatures for protection from energy development as well as the Congressional record of the debate that ensued because of that conflict. Overall, as ranching interests turned to federal and state legislatures for protection, they found that elected officials now favored the interests of energy development and that the existing legal and regulatory frameworks also favored energy development. As a result, the formerly allied interests of ranching and energy developers began to compete for the attention and favor of elected officials as each sought control of the BLM's land-use subgovernment.

Keywords Congress · Bureau of Land Management · Federal regulation · State legislature · Surface Owner Protection Act · Split-estate energy development

This chapter describes how changes in federal domestic energy policy resulted in increased split-estate energy development and an analysis of the effects of that development. Information for this chapter was collected from primary government documents and journalistic sources across three settings, New Mexico, Colorado, and Wyoming, where the vast majority of split-estate energy development and conflict between ranching and energy development interests have occurred. Analysis of the data shows increased split-estate energy development raised the level of frustration with the BLM among ranchers and a spiraling conflict via accounts of ranchers petitioning their state legislatures for protection from energy development. Overall, as ranching interests turned to state legislatures for protection, formerly allied interests increasingly competed for control of the BLM's land-use subgovernment.

© Springer Nature Switzerland AG 2019   51
R. E. Forbis Jr., *Altered Policy Landscapes*,
https://doi.org/10.1007/978-3-030-04774-0_4

Networks Within Subgovernments

Subgovernments are composed of three networks of actors linked by a shared inter-est in a policy domain (Cater, 1964; Davis, 2001; Freeman, 1965; Maas, 1949; Ripley & Franklin, 1984). The composition of these subgovernment networks has traditionally been described as an alliance between congressional committees, exec-utive departments or bureaus, and interest groups for the purpose of controlling policymaking decisions. Over time, these alliances have been reformed in a manner that allows for a broader array of interests to participate in the process of shaping policy decisions. Despite this fragmentation, subgovernments have retained the basic structure of three primary networks working in alliance to control the policy domain (Davis, 2001). A subgovernment retaining its basic structure means that units of stakeholders operating within the subgovernment network interact in response to fragmentation of the policymaking environment. As units of stakehold-ers respond to fragmentation, they compete for control of the policy environment. As stakeholders compete for dominance, a hierarchy is established within the net-works of the subgovernment. Competition means that power is then dispersed among the various units of policy stakeholders as one unit of stakeholders, or a combination of stakeholders, establishes domination of the policymaking environ-ment (McCool, 1987, 1989, 1990, 1995, 1998).

Stakeholders' response to fragmentation can range from cooperation to competi-tion. The choice to support greater cooperation or engage in competition is depen-dent on two factors: first, the choice to cooperate or compete is dependent upon whether or not fragmentation threatens to disrupt established network hierarchies within the policy subgovernment and, second, whether or not disruption to one net-work's hierarchy is supported by actors operating within the remaining two net-works. While none of the networks is immune to change, policymaking subgovernments as a whole are notable for their stability over time. Simply put, minor changes happen relatively often, but major disruptions occur infrequently: when a major disruption does occur, it is often the result of considerable effort to legislate sweeping reforms to an entrenched hierarchy within the subgovernment.

The vulnerability of established network hierarchies to change or disruption takes on a variety of forms. For example, instability in a network of congressional committees is most often the result of elections. Similarly, elections of a new execu-tive administration result in change to executive departments or bureaus. Nonetheless, electoral results overwhelmingly favor incumbents in elections, and executive departments or bureaus are noted for their ability to defend themselves from politi-cal interference. As one might expect, hierarchies within these particular networks have remained relatively stable over time. The same cannot be said, however, for the network hierarchies of interest groups.

The hierarchy of interest group networks is vulnerable to political disruption. This disruption is a result of the relative stability of hierarchies operating within the more politically oriented networks of congressional committees and administrative departments or bureaus. This is particularly true when political networks and their internal hierarchies share a common policy objective whose success is dependent upon their political intervention. In a sense, because these networks' hierarchies are dominated by elected officials who share a common objective, their strategic political action to achieve the desired objective is united. The effect of this political unity destabilizes the existing hierarchy within the remaining interest group-oriented network. For example, within the land-use decision-making subgovernment, if it is the favored policy objective of the political networks to expand domestic energy development, this form of political unity would then effectively threaten to disrupt ranching's long established dominance of the interest group network.

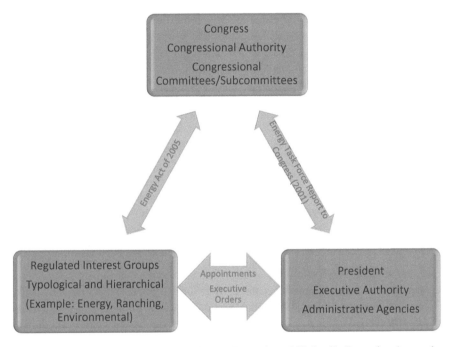

Subgovernment Theory Operationalized for the Expansion of Hydraulic Fracturing Across the States of the Rocky Mountain West

The political leadership of natural resource congressional committees shared the Bush-Cheney administration's objective of expanding domestic energy development. In turn, its decision-making activities were merged with those of the Bush-Cheney administration. United by a common objective, the collective political decision-making of elected officials instigated a major disruption to the hierarchical relationship between ranching and energy development. Disruption to this relationship then triggered conflict and competition between the formerly allied, strong, resource-rich members in the BLM's public land subgovernment: ranchers and energy developers.

What emerged from the disruption was a highly publicized political conflict between ranchers and energy developers. As these two interest groups were his-

torically supportive of one another, disruption at the hands of elected political
leaders was particularly troubling to each group. Ranching and energy developers
maneuvered for dominance of the land-use policy domain, and as they engaged, it
became clear that the long and convenient marriage of ranching and energy was
coming to an end.

Change Comes to Cowboy Country

Politically motivated changes in the BLM's domestic energy policies resulted in
increased split-estate energy development. To effected ranchers, the economic
losses they were suffering as a result of developing their lands and resources for
energy extraction did not compare favorably to the economic gains being derived by
a government-sponsored energy industry. As the regulatory disparities and economic inequities began to affect more and more ranchers across the West, the ranchers' frustration intensified. Ranchers' frustration was compounded by a perception
that the federal government was unwilling to consider legislative or regulatory
reform. It did not help matters that energy development interests were unwilling to
support ranching's efforts at reforming split-estate energy development (Clifford,
2001; Hardin & Jehl, 2002; Mitchell, 2005).

    Legally, the federal government has the right to access federally owned minerals
(Bureau of Land Management, 2007). That right is conveyed to private energy
development companies when companies purchase a federal mineral lease. The
right of the company to access the lease for development is implicit despite the fact
that an individual might own the property on which the mineral lease is located. In
essence, the land is split into two estates. One is known as the subsurface mineral
estate, and the other is known as the private surface estate. At its modern peak,
3–5% of all energy development in the Western states occurred on split-estate lands
(Environmental Working Group, 2004). While some split-estate development
occurred relatively peacefully, most split-estate development met considerable
opposition. Led primarily by Western ranchers whose lands were being leased
developed at what was, for them, an alarming rate (Hardin & Jehl, 2002). As part of
their opposition, ranchers first turned to the BLM for assistance in helping them to
understand why their lands were suddenly and swiftly being targeted for energy
development (Miller, Hamburger, & Cart, 2004). However, the BLM's ability to
assist ranchers was restrained by the legal and regulatory dominance of the mineral
estate. In the past, ranchers had worked amicably, and at their own pace, with energy
companies and the BLM in developing mineral leases (Mitchell, 2005). However,
spurred by their political masters, the BLM's effort to rush forward with industry
access and development of energy leases was quite suddenly overwhelming to
ranchers (Clifford, 2001).

Regulating Split-Estate Energy Development

Access to the surface estate for the development of the subsurface estate is regulated by Federal Onshore Order #1 (Bureau of Land Management, 2009; Parks, Forests, and Public Property, 2008; Public Lands: Interior. Minerals Management: General, 2008). Known simply as the "Gold Book," this body of formal rules and regulations, operating standards, and best management practices, last revised in 2007, requires that the surface owner be notified of the company's intent to explore and develop the mineral lease prior to accessing the private property. The regulations also require that, upon acknowledgment of having received notification of intent to explore and develop the mineral lease, the company and the property owner should negotiate the terms of access as well as any development activities that may take place. These contracts are known as "surface-use agreements" (Bureau of Land Management, 2007).

Negotiation of surface-use agreements is unregulated by federal or state administrative agencies. Surface-use agreements are considered private contracts. The terms and conditions of these agreements are negotiated between the energy lease developer's representative—known simply as a "land man"—and the surface property owner. Once negotiations are concluded, federal regulations require that a certified notification with the BLM be filed to show a surface-use agreement has been reached and when development activities are scheduled to begin. Governmental oversight of the company's exploration and development activities is on the private surface, unless otherwise noted in the contract, the sole responsibility of the property owner (Bureau of Land Management, 2007).

> The surface use agreement between the surface owner and the operator is confidential. However, the APD Surface Use Plan of Operations must contain sufficient detail about any aspects of the agreement necessary for NEPA documentation and to determine that the operations will be in compliance with laws, regulations, Onshore Orders, and agency policies. When the operator submits its Surface Use Plan of Operations to the BLM, the operator must make a good faith effort to provide a copy to the surface owner. Following APD approval, the operator must also provide a copy of the Conditions of Approval to the surface owner. In addition, the operator must make a good faith effort to provide a copy of any proposal involving new surface disturbance to the private surface owner. (43 C.F.R 3104 and 36 C.F.R. 228 Subpart E as cited in Bureau of Land Management, 2007, p. 12)

The legal precept known as "liberty of contract" guides surface-use negotiation and agreement. Liberty of contract is a free-market principle where both parties enter into negotiations free from government interference for the purpose of entering into legally binding contracts (Buckley, 1999; Fitzgerald, 2008, 2009). Fundamentally, the surface owner and industry representative negotiate in a manner that seeks to protect their respective self-interests free from government intervention. To some, freely negotiating terms of access and development, from the standpoint of protecting one's self-interest, is the preferred process (Buckley, 1999; Fitzgerald, 2008, 2009). However, self-interests aside, these types of contracts essentially absolve the BLM from regulating what type or form of compensation and/or mitigation should or should not be addressed in the surface-use agreement. This means that unless surface owners have sufficient knowledge of federal and state regulations of energy development processes or competent legal counsel, split-estate property owners are left to their own devises in negotiating the surface-use agreement.

> The operator must make a good faith effort to notify the private surface owner before entering private surface to stake a well location and access road or to conduct cultural or biological surveys. The BLM will invite the surface owner to participate in the onsite and final reclamation inspections and will take into consideration the needs of the surface owner when reviewing the APD and reclamation plans and when approving final abandonment and reclamation. The BLM will offer the surface owner the same level of surface protection that the BLM provides on Federal surface. The BLM will not apply standards or conditions that exceed those that would normally be applied to Federal surface, even when requested by the surface owner. (43 CFR 3104 and 36 CFR 228 Subpart E as cited in Bureau of Land Management, 2007, p. 12)

There are currently three different surface-use agreement contract forms recommended for use by energy developers and surface owners (Western Governors' Association, 2004). That there are so few, and that they vary so greatly in what is recommended for negotiation, is an example of the unregulated nature of surface-use negotiations and agreements. No statutory standard exists governing surface-use agreements. This means that surface owners negotiate surface-use agreements from a position of unequal footing. The potential risk is that serious harm befalls a surface owner's economic and environmental stewardship. Additionally, because terms of the surface-use agreement are undefined by federal or state regulations—and remain ill-defined by existing statutory language—there is substantial confusion over the exact nature and scope of economic losses from surface energy development activity.

> The operator must negotiate in good faith with the surface owner. Negotiating in good faith provides a forum through which the operator and surface owner can discuss the preferences and needs of both the surface owner and the operator. In addressing those needs, the operator may be able to modify the development proposal to both minimize damage to the surface owner's property while reducing reclamation and surface damage costs. For example, operator costs can-might be [sic] minimized by placing roads and facilities in locations that meet the surface owner's long-term development plans for the property, thereby lessening the future reclamation obligations of the operator. (43 CFR 3104 and 36 CFR 228 Subpart E as cited in Bureau of Land Management, 2007, p. 12)

In combination, lack of regulation, unequal footing, and confusion create tension between surface owners (ranchers, farmers, and homeowners), subsurface developers (energy companies), and the federal government (BLM).

The lack of a surface-use agreement does not mean that access to the surface by the leaseholder can be denied by the surface owner. Access to the surface estate for the purpose of developing the energy lease cannot be denied. If an agreement for access and development cannot be reached, an appeal is filed with the BLM by the leaseholder. If the appeal is upheld, the developer is then required to post a bond to financially compensate for any foreseeable damages that may occur during the course of development and/or to cover the costs of surface reclamation after development is concluded.

Prior to approval of the APD (or Sundry Notice to conduct new surface disturbing activities), the operator must certify as part of the complete application that a good faith effort had been made to reach a surface use agreement with the private surface owner and that an agreement was reached or that it failed. If the surface owner and operator fail to reach an agreement, the operator must file a bond with the BLM ($1000 minimum) for the benefit of the surface owner to cover compensation, such as for reasonable and foreseeable loss of crops and damages to tangible improvements. Prior to approving the APD, the BLM will advise the surface owner of the right to object to the sufficiency of the bond and will review the value of the bond if the surface owner objects. The BLM will either confirm the current bond amount or establish a new amount. Once the operator has filed an adequate bond, the BLM may approve the APD. Following APD approval, the operator and the surface owner may appeal the BLM's final decision on the bond amount. (Oil and Gas Leasing, 1988; Parks, Forests, and Public Property, 2008 as cited in Bureau of Land Management, 2007, p. 12)

If a conflict between the surface owner and leaseholder does occur prior to an agreement being reached, or during the course of the lease's development—the cause of which can take on multiple forms—a complaint is filed with the BLM requesting the agency's intervention and assistance in resolving the issue (Clifford, 2001; Mitchell, 2005). Thus, during the course of negotiating terms of surface-use agreements, the parties negotiate with the knowledge that access and development cannot be denied and that any denial by the surface owner will be met with government intervention.

Regulatory authority of federal energy leases is delegated to the BLM. In turn, the BLM has, over time, promulgated regulations ensuring the federal subsurface remains accessible to the government's development agents for the purpose of bringing energy resources to market. Thus, an energy company's vested property right is not simply government's enforcement of access to the surface, but it is also the result of government protecting the energy lease's economic development. As discussed in Chap. 2, governmental property rights to the mineral subsurface are a product of the late nineteenth-century and early twentieth-century homesteading, mining, and energy legislation. The legislation, therefore, conveys a property right with industry's purchase of an energy lease, developing the energy resource and bringing it to market. In essence, with the purchase and development of a federal energy lease, the property right of access and economic benefit of ownership are conveyed to energy companies.

Thus, energy development regulations—Onshore Order #1—sustain property rights and economic interests to multiple parties in the use and development of a split-estate. Access and development of valued resources within the boundaries of a split-estate property are essentially shared between surface owner, lease holder, and federal government. As one might surmise, these rights and interests are a tangled web. The entanglement of these rights and interests is a result of general provisions in Title 43, Chap. 7, Subchapter X, at Statute 299 entitled "Reservation of Coal and Mineral Rights" of the Stock Raising Homestead Act of 1916b (SRHA):

All entries made and patents issued under the provisions of this subchapter shall be subject to and contain a reservation to the United States of all the coal and other minerals in the lands so entered and patented, together with the right to prospect for, mine, and remove the same. The coal and other mineral deposits in such lands shall be subject to disposal by the United States in accordance with the provisions of the coal and mineral land laws in force at the time of such disposal. Any person qualified to locate and enter the coal or other mineral deposits, or having the right to mine and remove the same under the laws of the United States, shall have the right at all times to enter upon the lands entered or patented, as provided by this subchapter, for the purpose of prospecting for coal or other mineral therein, provided he shall not injure, damage, or destroy the permanent improvements of the entryman or patentee, and shall be liable to and shall compensate the entryman or patentee for all damages to the crops on such lands by reason of such prospecting. Any person who has acquired from the United States the coal or other mineral deposits in any such land, or the right to mine and remove the same, may reenter and occupy so much of the surface thereof as may be required for all purposes reasonably incident to the mining or removal of the coal or other minerals.... (Stock Raising Homestead Act of 1916b, Title 43,299: Reservation of Coal and Mineral Rights)

The entanglement of agricultural property and energy economics is an ill-suited statutory vehicle for the development of energy resources in the twenty-first century. Regulations are ill-suited because they do not seek to balance the rights and interests of the government's agent with those of the surface owner. This is due to the statutory language that specifically limits compensatory damages to the surface estate resulting from energy development activities. Consequently, modern energy development's regulated compensatory responsibilities remain minimal at best. The provision, "Reservation of Coal and Mineral Rights," also establishes minimal compensation for losses suffered by owners of split-estate surface lands and resources:

...first, upon securing the written consent or waiver of the homestead entryman or patentee; second, upon payment of the damages to crops or other tangible improvements to the owner thereof, where agreement may be had as to the amount thereof; or, third, in lieu of either of the foregoing provisions, upon the execution of a good and sufficient bond or undertaking to the United States for the use and benefit of the entryman or owner of the land, to secure the payment of such damages to the crops or tangible improvements of the entryman or owner, as may be determined and fixed in an action brought upon the bond or undertaking in a court of competent jurisdiction against the principal and sureties thereon, such bond or undertaking to be in form and in accordance with rules and regulations prescribed by the Secretary of the Interior and to be filed with and approved by the officer designated by the Secretary of the Interior of the local land office of the district wherein the land is situate, subject to appeal to the Secretary of the Interior or such officer as he may designate.... (Stock Raising Homestead Act of 1916a, Title 30,54: Liability for Damages to Stock Raising and Homestead Entries by Mining Activities).

This means that neither the federal government nor its development agent—energy companies—has legal responsibility to compensate for economic losses beyond those that result in damage to crops or existing tangible improvements. Any additional compensation is an instrument of the negotiated surface-use agreement or an appeal process filed through the BLM. As part of the appeal process, there is a statutory requirement of a financial bond being secured against the potential for economic losses resulting from damage to the surface and the cost of reclamation. This process is commonly referred to as the practice of "bonding on." Thus, in the bond, access to the surface for the benefit of developing the energy resource is, once again, secured via government intervention. The legislative intent for those requirements, and subsequent administrative interpretation, is found in the statutory language of the Stock Raising Homestead Act of 1916a, 1916b (SRHA)and the Mineral Leasing Act of 1920 (MLA) (revised in 2001).

The amount of monies required for these bonds varies and is defined by federal regulation—Onshore Order #1. Essentially, the amount of the bond is dependent on the level of energy development being proposed as well as the BLM's interpretation of legislative intent for the purpose of requiring the bonds. Under the SRHA, the bond must exceed $1000 and is intended to recover potential damages to crops or tangible improvements existing on the surface at the time of development (Stock Raising Homestead Act of 1916a, 1916b). However, this leaves open to administrative interpretation compensation for potential loss of income or economic benefit in any future use of the surface and its surrounding resources. Further, according to the revised MLA, a developer must secure a bond in the amount of at least $10,000 per lease to ensure compliance with environmental protection measures (Mineral Leasing Act of 1920, amended 1987; Oil and Gas Leasing, 1988; Parks, Forests, and Public Property, 2008 as cited in Bureau of Land Management, 2007, p. 13). Bonds of this type are the result of modern environmental and mining reclamation legislation such as the National Environmental Protection Act of 1969 (NEPA) and Surface Mining Control and Reclamation Act of 1977 (SMCRA) (National Environmental Protection Act of 1969; Surface Mining and Control and Reclamation Act of 1976, amended 1993; Department of Interior: Office of Surface Mining Reclamation and Enforcement, n.d.). Administrative responsibility of these types of environmentally oriented bonding requirements is delegated to the BLM in keeping with the Federal Lands Policy and Management Act of 1976 (FLPMA). Federal energy bonding regulations allow for literally hundreds of oil and gas wells being drilled on just one lease, or in one state, or for that matter, across multiple states within a region containing energy resources.

Federal regulations allow energy developers to post what is known as a "blanket bond" (Mineral Leasing Act of 1920; Lease of Oil and Gas Lands, 1988). Blanket bonds of up to $25,000 are required for all wells an energy company might drill in one state. A company securing a bond in the amount of $150,000 allows energy developers to operate in more than one state regardless of the number of wells it expects to drill.

> The bond may be a surety bond or pledge backed by cash, negotiable securities, Certificate of Deposit, or Letter of Credit in the minimum amount of $10,000. In lieu of a $10,000 lease bond, a bond of not less than $25,000 for statewide operations or $150,000 for nationwide operations may be furnished. (Lease of Oil and Gas Lands, 1988; Parks, Forests, and Public Property, 2008 as cited in Bureau of Land Management, 2007, p. 13)

Given the expected return from any one producing well—estimated at $20 Million—these bond requirements are relatively easy to secure and, as some have noted, do not begin to cover potential environmental damage and economic loss that might result from energy development activities on split-estate lands (Sievers, 2004). Therefore, while access and development are shared between stakeholders in a split-estate, there is significant political and regulatory disparity in the stakeholders' ability to derive economic benefit from their shared use of a split-estate. These disparities helped establish the conditions for a political conflict to emerge between ranchers and energy developers.

Energy Politics and Policy: Congressional Committees (2000–2008)

From 2001 to 2008, numerous congressional committee meetings were held to address energy policy. The topic of most of these energy-related hearings was focused on the nexus of national security and the programmatic expansion in developing domestic reserves. While hearings were convened to address a wide range of topics, hearings held to address problems associated with the expansion of energy development were few. The record of committee hearings indicates that when complaints were heard, the testimony of ranchers was often included, but the focus of the inquiry was more concerned with removing regulatory roadblocks to expand domestic energy development (Congressional Hearings, 107th–110th Congresses). Thus, voices representing ranching operations impacted by the rapid expansion of domestic energy development were secondary to voices representing the interests of energy developers.

The testimony of ranchers often followed statements from the committee's chair expounding the virtues of expanding domestic energy development. Or, as was often the case, ranchers' testimony preceded the testimony of numerous energy spokespersons. A review of the record of House Natural Resource Committee and Subcommittee hearings between 2000 and 2008 clearly indicates that testimony from ranching interests was wedged between articulations of political support from elected officials and the policy recommendations of energy representatives (Congressional Hearings, 107th–110th Congresses). While lone ranchers spoke on behalf of ranchers besieged by energy development, elected representatives, state officials, energy scientists, and members of the energy lobby spoke on behalf of speeding up the regulatory permitting process or expanding energy leasing sales (Hearing on enhancing America's energy security, 2003; Oversight Hearing on the orderly development of coalbed methane resources from public lands, 2001). For example, during the 107th Congress, in an oversight hearing before the House Energy and Mineral Resources Subcommittee, its subcommittee chair, Barbara Cubin (R-WY), described the effect of split-estate energy development as "unconventional" and the effects of split-estate development as "growing pains."

> As with any resource, such an explosion of activity comes with "growing pains" while individuals, communities, local and state government and public land managers attempt to plan for the costs and benefits associated with the extraordinary interest in CBM…Split-estate mineral development is often contentious - and when conflicts arise they grab the headlines. Steady royalty income to a fee mineral owner happy with his check is a "dog

bites man" story. When a rancher gets cross-wise with a driller seeking to access his federal lease, or other fee mineral ownership from which the rancher does not financially benefit, then that becomes a "man bites dog" story. When a lot of ranchers without minerals get upset, that's a [c]over story in Time Magazine…eastern media reporters have written tales of ranchers with new pick-ups paid for by CBM royalties, followed by tales of grazing lands ruined by the unregulated discharge of produced waters. On top of this are stories that Montana and Wyoming governments are "at war" with one another over surface water quality…Well, I live out there, and if there is a war going on, it's about the federal government getting sufficient funding for the Bureau of Land Management to complete a cumulative impacts analysis of anticipated CBM development so that land-use plans can be updated, and mitigating measures drawn up, to allow federal lessees to drill and bring their gas to market… the real question is "how can we best mitigate these conflicts?" Do ranchers need a "surface owners" Bill of Rights", and if so, which level of government ought to be considering it? On the other hand, when surface owners acquired the title to their property did they not understand what it meant to have mineral rights reserved to the government or another individual? (Oversight Hearing on the orderly development of coalbed methane resources from public lands, 2001)

To further the indication of unequal footing of ranching interests, congressional hearings to consider the administration's proposal to expand domestic energy development were held before committees chaired by political allies from the energy-producing Western states (Congressional Hearings, 107th–110th Congresses). Thus, the merits of the proposed expansion to domestic energy development as outlined in the report of the President's Energy Task Force met with considerable political favor. This was particularly true of committees whose oversight responsibilities concerned the administration of public lands and resources.

Republican dominance of congressional committees helped the Bush-Cheney administration achieve the objective of expanding domestic energy resource development. In part, this is because the 2000 presidential election marked the return of Republican control to both houses of Congress. This meant that political control of congressional public lands and resource committees were dominated by members of the president's own political party (Congressional Hearings, 107th–110th Congresses). And, once again, the events of September 11, 2001, would provide much needed justification and public support to partisan committee chairs and members as they acted to support and enact the administration's energy policy master plan. For instance, a quick survey of congressional hearings held during the 107th Congress (2001–2003) shows that roughly 30 hearings have been devoted to deliberations of energy policy in the context of national security (Congressional Hearings, 107th–110th Congresses).

At the time of the 107th Congress, there was near unanimity among Western states' congressional delegations in support of expanding domestic energy development (Congressional Hearings, 107th–110th Congresses). Their support, however, was only partially ideological in nature. If the rationale of self-interest among elected officials was true, then support was primarily a result of political reality in their desire for reelection (Douglas, 1990; Edelman, 1988; Habermas, 1975; Kelman, 1987; Kingdon, 2003; Levine & Forrence, 1990; Offe, 1985). This is because, as elected representatives from the energy-producing states of the West, they were very cognizant of the economic benefits that result from increasing energy development in their home states.

The BLM is mandated by the Congress to hold quarterly energy lease auctions (Competitive Leases, 1988). Monies from federal sales of these energy leases and royalties from the energy's development are shared with the states. Thirty-five percent of monies collected from these auctions go directly to the state where the energy leases are located (Oil and Gas Royalty, 1988). Thus, every 3 months in each of the energy-producing states of the Rocky Mountain West, energy leases are auctioned to the highest bidder. However, some energy-producing states, like Wyoming and Montana, regularly offer energy lease sales on a bimonthly basis (Bureau of Land Management, n.d.). Once the development of the lease occurs, the royalty from the fluid energy mineral produced is a 50–50% split between the federal government and the state (Bureau of Land Management, n.d.). In the rush to extract domestic energy resources, these financial incentives proved beneficial in industry's ability to achieve federal and state support for expanding their development activities.

Disruption, Conflict, and Competition: Energy and Ranching

The process of developing split-estate energy resources is disruptive. The process, even when performed properly, negatively impacts the working environment of most ranching operations in the West. The process of drilling for and extracting energy resources, particularly CBM, "can turn ranches and prairies into sprawling industrial zones, laced with wells, access roads, power lines, compressor stations and wastewater pits" (Hardin & Jehl, 2002). The long-term impact of these extractive processes can be debilitating to surface owners.

> …the artesian well on Roland and Beverly Landrey's ranch has failed. After producing 50 gallons a minute for 34 years, the well, the ranch's only source of water, stopped flowing in September. A well digger who examined it blames energy companies drilling for gas nearby, but the companies dispute that. So the couple—he is 83 and ailing; she describes herself as "no spring chicken"—hauls water in gallon jugs and rives 30 miles to town weekly to wash clothes and bathe…Dave Bullach, a welder who lives near Gillette, couldn't take it anymore. For two sleep-deprived years, he endured the incessant yowl of a methane compressor, a giant pump that squeezes methane into an underground pipeline. There are thousands of these screaming machines in Wyoming, where neither state nor federal law regulates their noise. Mr. Bullach stormed out of his house at midnight last year with a rifle and shot at the compressor until a sheriff's deputy hauled him off to jail. (Hardin & Jehl, 2002)

Energy resources, cheap energy resources, like that of CBM, had become increasingly feasible for industry to extract, develop, and market. This is because there had been substantial and important breakthroughs in energy technology. A process known as hydraulic fracturing, or "fracking," where chemically treated water is forced into tight seams of coal formations in the effort to loosen the methane gas for collection had been perfected (US Environmental Protection Agency, 2000). The engineering feat of being able to collect and capture the methane gas from multiple points at a single location, a technique known as "directional drilling," had also been perfected (Kennedy, 2000).[1] Furthermore, fracking and directional drilling emerged just prior to the Bush-Cheney administration taking office. In their infancy neither the process of fracking nor the technique of directional drilling was widely used by industry; both were considered cost-prohibitive. But by 2001, the cost of energy resources raised as rapidly and as steadily as the energy-friendly political decisions being made by the Bush-Cheney administration and Congress. In turn, fracking and directional drilling became cost-effective. While these new means of extracting hard-to-get energy resources are cost-effective and efficient, the process of fracking is problematic.

> As it runs through Orin Edwards's ranch, the Belle Fourche River bubbles like Champagne. The bubbles can burn. They are methane, also called natural gas, the fuel that heats 59 million American homes. Mr. Edwards noticed the bubbles two years ago, after gas wells were drilled on his land. The company that drilled the wells denies responsibility for the flammable river. (Kennedy, 2000)

Most CBM energy resources lie within very tight, close-to-the-surface seams of coal. This fact is one reason why states of the Rocky Mountain West experience the largess of the modern energy boom: its benefits as well as its problems. One problem with the process of fracking is its effect on the water resources of a state, a community, a subdivision, or a ranch. The water required for the CBM fracking process varies depending a number of factors, including the depth and type of coal seam formation being utilized. Nonetheless, in shallow seams, like those found in the Powder River Basin of Northeast Wyoming, a typical CBM well will use 400 barrels (42 gallons/barrel) of water per day (16,800 gallons/day) (United States Geological Service (USGS), 2000).

Throughout the CBM-producing states of the West, energy development's use of water is a contentious issue for ranchers. One 2002 estimate expected that in the Powder River Basin alone the energy industry would "pump out 3.2 million acre feet of water—as much as New York City uses in two and a half years" (Hardin & Jehl, 2002). This water's use is limited. For example, when treated properly, the extracted water can be beneficial to ranchers. However, when not treated properly, much of the water is riddled with saline which, if untreated and dispersed over pasture lands, can turn grazing lands into barren wastelands (Clifford, 2001; Hardin & Jehl, 2002; Mitchell, 2005). To make matters worse, wastewater is disposed by reinjection or spraying across pasture lands, a common practice among energy developers. Additionally, most wells or clusters of wells produce far greater amounts of water that any one rancher can use. Water use aside, the fracking of the coal seam has the attendant effect of releasing un-captured methane gas and transferring it to

---

[1] Note: Kennedy is commenting on Summary and Analysis of Department of Energy Office of Fossil Energy Reports concerning the advancement of directional/horizontal drilling technologies.

free-flowing water sources like irrigation streams or water wells. Thus, surface property owners with champagne-like irrigation streams make a habit of documenting the effect of the fracking process by taking matches and lighting the bubbles on fire (Anderson, 2009).

Water is not the only impact to surface owners from the process of extracting CBM energy resources. The effect that energy development can have on a surface owner's property interests ranges from a simple nuisance like dust to depleting a water aquifer to the point where water pumps burn out and fail (Clifford, 2001; Hardin & Jehl, 2002; Mitchell, 2005). Ranchers in particular bear the brunt of multiple impacts that disrupt their stock raising operations: cattle and sheep killed by energy traffic, chemical spills from poorly constructed drill holes, as well as erosion from newly cut and heavily traveled roads, pipelines cutting across grazing lands, and drilling pads dotting the land (Clifford, 2001; Earthworks, n.d.-b; Hardin & Jehl, 2002; Mitchell, 2005).

As split-estate energy development expanded and the problems became more widespread, the frustration of ranchers began to escalate. In turn, the traditionally friendly communication between ranchers and the energy industry began to rapidly deteriorate. Battle lines between these two interest groups were beginning to form. As representatives sought to alleviate the anger of ranchers, the conflict grew evermore heated. For example, in Wyoming, a group known as the Coalbed Methane Coordination Coalition failed to keep the peace between ranching and energy producers. When the energy industry charged the coalition with being too sympathetic to ranching and environmental interests, it stopped funding the coalition. While the coalition still exists, its director notes that "polarization and demonization are absolute hallmarks of drilling for coal-bed methane" (Hardin & Jehl, 2002). While the severity of rancher's anger is attributable to split-estate energy development, that anger reached its tipping point as the BLM began "reducing the number of cattle it allows to graze on federal lands" (Hardin & Jehl, 2002). The likelihood of balancing the interests between the ranchers and energy producers, as had been common in the past, is aptly described by the BLM New Mexico's Steve Henke, "Ranchers are losing out to the energy industry in terms of their capability to grow grass... Stepping back though, what's in the public interest? It's not that this area is unsuited to ranching. But we've got a world-class gas resource here" (Hardin & Jehl, 2002). To effected ranchers, it seemed as though the BLM had turned a deaf ear to their complaints. However, in defense of the BLM, agency administrators were limited in their capacity to offer affected ranchers mitigation and remediation. This is because a shift in BLM policy and resources had occurred. BLM's new land-use policy emphasis was on increasing domestic energy development. Simply put, the BLM was no longer primarily concerned with grazing or appeasing ranchers (Wilkinson, 2005).

The unfettered pace of energy development in the West occurred on both public and private lands, and by 2003, energy development dominated Western landscapes. In the San Juan Basin of New Mexico alone, 19,000 producing CBM wells dotted the open terrain where herds of grazing cattle had once roamed (Snell, 2003). In the rush to develop America's domestic energy resources, pastoral scenes of grazing cattle on the public domain had been replaced with the urban-like hustle and bustle of energy development. The sights and sounds of energy development were unsettling to those who had grown accustomed to serenity. Where ranchers had just a year earlier grazed their herds, an infrastructure of active drilling rigs, thousands of concrete well-pads, miles of pipelines, and tens of thousands of miles of roadways, where hundreds of vehicles—large and small—rumbled throughout the day and the night, had taken their place. A modern-day tragedy of the commons in the form of energy development was beginning to unfold across the Rocky Mountain West.

Unlike the earlier tragedy of unfettered grazing, however, energy resource development had not been confined to the public domain. The effect on ranching operations, some of which had been in existence for generations, elicited fierce responses among members of the ranching community.

> …ranchers like Velasquez—fiercely independent, sometimes cantankerous, and almost always politically conservative—are beginning to organize and fight back. Last year, for example, Velasquez and several other ranchers got so fed up with what they see as oil and gas development run amok that they locked the gates to their private land…The [energy] companies called official at the BLM; it was clear a rebellion was taking shape. (Snell, 2003)

To affected ranchers, the impact of energy development went far beyond their loss of peace and quiet. Ranchers' list of grievances, like their anger and frustration, grew with each new well being drilled. Their grievances include: Toxic chemical spills, ranch gates being left open, grazing lands scraped in one to six acres patches at a time, soil erosion from roads, neglected reclamation, and the introduction of invasive weed infestations, and, worst of all, the death of cattle. The frustration and anger of ranchers, farmers, and homeowners affected by split-estate energy development were palpable.

In turn, ranchers sought the help of their traditional advocacy groups such as their local Cattlemen, Stockgrower, and/or Woolgrower Associations and Farm Bureaus. What ranchers soon realized, however, was that while their traditional associations were sympathetic to their plight, association's advocacy on behalf of members seeking assistance was moderated by the associations' historical cooperation with energy developers (Royster, 2004). As a result, affected ranchers, farmers, and home owners began to form non-traditional advocacy organizations to confront the problems associated with split-estate energy development. Organizations such as the San Juan Citizens Alliance of New Mexico, the Powder River Basin Resource Council of Wyoming, the North Fork Ranch Landowners' Association of Colorado, and the Landowners' Association of Wyoming began to appear across the West. In essence, ranchers began forming non-traditional advocacy organizations with the purpose of lobbying elected officials for assistance and reform of existing laws and regulations guiding split-estate energy development.

Mending Fences? Western States' Surface Owner Protection Acts

Not long after ranchers organized into their newly created advocacy groups, they sought state legislative relief from the impact of federal split-estate energy development. Ranchers did so by petitioning their respective state elected officials. And as tensions rose and tempers flared, Western state legislatures began considering surface damage statues in the form of Surface Owner Protection Acts (Earthworks, n.d.-a).[2] The central concern of the legislation is to mitigate surface damage to private lands while still allowing for responsible energy development.

Generally speaking, most of this type of legislation had been passed in response to the previous domestic energy boom of the 1970s and 1980s (Earthworks, n.d.-a). However, in the energy boom of the 1970s and 1980s, most Western states' energy development had been mostly confined to public lands. Thus, by 2000, no Western state other than Montana had legislated statutory relief to offset private surface damage or economic losses incurred by split-estate energy development. In the modern energy boom of the late 1990s and 2000s, as the federal government pursued a course of expanding domestic energy development, the states of New Mexico, Colorado, and Wyoming struggled to pass their version of surface owner protection legislation (Colorado Surface Owner Protection Act, 2007 as cited in Earthworks, n.d.-a; New Mexico Surface Owner Protection Act, 2007; Wyoming Surface Owner Accommodation Act, 2005).

The newly formed landowner associations thought state protection of their surface lands and resources was a reasonable request to make of their state elected officials. However, state energy associations viewed state protection of surface estates a form of unwarranted governmental interference that circumvented federal law, as well as being economically burdensome (Associated Press, 2005a; Bleizeffer, 2004a, 2004b, 2004c; Farquhar, 2002). Energy development's opposition was viewed as a bit puzzling to surface-owning ranchers because most surface damage statutes are incredibly similar in that they provide state protection from unreasonable damages to the surface estates of private landowners. Surface owner protection acts serve a threefold purpose: (1) to minimize harm to individual surface estate owners affected by development of the mineral estate; (2) to minimize harm to the general public suffered when agricultural lands, or public lands, are damaged by the development of the mineral estate; and (3) to foster the reasonable development of the mineral estate through the prevention of unsettled disputes between surface and mineral estate owners (Alspach, 2002).

Surface owner protection acts have the effect of altering the traditional relationship between the surface estate owner and the mineral estate developer (Walker, 1983). On the one hand, surface owner protection acts are beneficial as they often extend notification periods and increase liability for damage to the surface estate in

---

[2] Note: Surface owner protection acts are not uncommon to states where energy development is a major economic activity. For example, prior to 2000, the states of North Dakota, Oklahoma, Montana, South Dakota, West Virginia, Tennessee, Illinois, Indiana, and Kentucky had all passed Surface Owner Protection Acts in one form or another.

excess of existing federal regulation. On the other hand, because these state actions substantially change the negotiating relationship between surface and subsurface estate owners, protective legislation can have a detrimental effect on energy developers. This is because requiring developers to address and account for potential harm specified by state legislation is considered to have the effect of "calling into question the fiscal advisability of oil and gas development" (Evans, 1996, p 515). While extended deadlines affecting surface owners are fairly common, the vast majority of surface owner protection acts specifically prohibit damage assessments based on any speculation of real estate value beyond established market price (Earthworks, n.d.-a). In their opposition to propose surface owner protection acts, representatives of energy producers confronted state lawmakers with the argument that the legislative impact would have a chilling economic effect on their ability to operate (Colorado Surface Owner Protection Act, 2007 as cited in Earthworks, n.d.-a; New Mexico Surface Owner Protection Act, 2007; Wyoming Surface Owner Accommodation Act, 2005).

The majority of this type of legislation contains statements of legislative purpose indicating that the legislation exists in order to further establish and foresee state interests.[3] Thus, there is state interest in defending ranching operations as an established economic activity and a state interest in fostering an environment where future energy development activity can take place. The legislative purpose sections of New Mexico, Colorado, and Wyoming versions of surface owner protection acts also identify the use of state police power to protect the public's environmental welfare (Earthworks, n.d.-a). Therefore, the energy development activity that does take place is performed in an environmentally sound manner. Moreover, all Western state surface owner protection acts cite both the protection of economic interests of its ranching and farming communities and ensuring compensation for those surface owners injured by mineral development (Earthworks, n.d.-a). Finally, the legislation also contains expressions of the state's desire to foster a peaceful coexistence between oil and gas developers and surface-owning citizens (Earthworks, n.d.-a). In doing this, states employ surface owner protection acts as means to protect the economic interests of both surface owners and energy developers, while regulating the environmental impacts from commercial energy resource development within their borders.

While other state's surface damage statutes have faced judicial scrutiny,[4] there has only been one test of the newly acted Western states' surface owner protection

---

[3] Note: There is no statement of legislative purpose included in the surface damage statutes of Indiana, Kentucky, Illinois, and Oklahoma.

[4] Note: The newly enacted surface owner protection statutes of the West are similar in construction to other states' legislation. Therefore, to landowners advocating for state intervention on their behalf, there was every expectation to believe the newly enacted legislation would survive constitutional challenges. This is because in previous challenges, both state and federal appellate courts had declared surface damage statutes to be constitutional exercises of state powers. For example, the Eighth Circuit Court of Appeals, addressing a due process challenge to North Dakota's surface damage act, declared the statute a constitutionally permissible exercise of state police power, in that the legislative protection of the state's agricultural and economic well-being, as outlined in the act, is substantially related to legitimate state interests. Similarly, the Oklahoma Supreme Court rejected the assertion that the Oklahoma surface damage act was an arbitrary and capricious exercise of the state's regulation of the public welfare. The court reasoned that, in passing its surface damage act, the Oklahoma Legislature sought to balance the rights of surface owners with those of mineral owners and that Oklahoma's act declares that surface lands are a resource as vital to the public welfare as the minerals beneath the surface.

acts: Wyoming's. The test came months after Wyoming's becoming the first among the modern energy-producing states of the Rocky Mountain West to pass surface owner legislation (Associated Press, 2005b; Bleizeffer, 2005; "The split-estate", 2006). When surface-use negotiations between a rancher and developer broke down over a request by the rancher for a reclamation bond of roughly $100,000, as per the new law, an appeal was filed with the state Oil and Gas Conservation Commission. The lease's developer claimed the amount being requested by the rancher was "off base" and that his posted state bond of $500, in addition to his other posted bonds with governmental entities was "sufficient." No matter the amount of the bonds, the energy developer claimed that reclamation had and always would be taken care of no matter the amount of the bonds his company had posted. The rancher disagreed and filed the appeal with the state commission. After hearing from the party's attorneys, the commission tossed the conflict back to rancher and developer, encouraging them to resolve the issue of the bond's adequacy. A year later, the matter was resolved having been negotiated to a resolution by the two parties. Neither rancher nor developer believed the new legislation had the intended effect of protecting their respective interests. The rancher believed that the new law did not help him much. The developer believed the new law only benefited lawyers (Associated Press, 2005b; Bleizeffer, 2005; "The split-estate", 2006).

State action regarding split-estate energy development remains somewhat limited. Not surprisingly, states' surface damage statutes, and their modifications to common law relationships, met with opposition from organizations representing energy development interests. The interaction between newly created statutory remedies and preexisting common law remedies has been termed a "somewhat unholy alliance" (Keffer, 1994, p. 525). The primary purpose of the typical state surface damage act is to provide affected surface estate owners with the financial means to restore their surface estate to its original condition or a condition as similar as possible to its premineral extraction condition. Unfortunately, the requirement of private negotiations between surface estate owner and mineral estate owner undermines this goal. The prospect of complex and expensive litigation results in an inequality in bargaining power between surface estate owner and mineral estate owner (Hageman, 1993, p. 292). Surface estate owners such as individual ranchers or farmers, in other words, are typically less able to bear the economic costs of litigation than corporate oil and gas companies. The mineral extraction industry, thus, has little incentive to come to a non-litigated agreement with a surface owner if the agreement is not substantially beneficial to their interests.

Conclusion

The inability to reform federal legislation and ill-defined regulations is one of the most troubling aspects of modern energy resource development. Specifically, reforming split-estate energy resource development is particularly difficult because the allied interests of rsanching and energy are firmly entrenched within the BLM's land-use subgovernment. Their policy partnership had grown so strong over time that each interest had effectively entrusted their respective future to the other. Within the interest group network of the BLM's land-use subgovernment, the fate and future of ranching and energy development had become inseparable. As a result, any attempt by ranchers to reform split-estate energy development threatened the interests of energy developers.

In the fight to reform split-estate energy development, energy interests have the legal and political upper hand. Because law and politics favor energy development, if ranching is ever to realize its desired reform of split-estate energy development, ranching organizations are going to have to negotiate with their industry counterparts. Or, alternatively, they must engage in full-blown political conflict if they are to secure their interests.

Given the long history of ranching and energy development's working relationship, negotiation would seem, at first glance, the likely pathway to reform. However, it is quickly apparent that energy is hesitant, even unwilling, to negotiate with ranchers or ranching organizations as they had once before. This is because the energy industry is secure in the knowledge that it has the legal upper hand as well as the support of elected government officials. In essence, energy developers have won over the very same government decision-makers that ranchers had themselves once enjoyed. To say that this is particularly galling to ranchers would be an understatement. As ranchers' attempted conciliatory negotiation with energy and engage in political outreach to elected officials, their efforts were repeatedly thwarted. In turn, their collective anger steadily increased and their political conflict with the energy industry escalated.

As was discussed in previous chapters, the BLM, guided by existing laws, regulations, the political mandate of a new administration, and the support of congressional committees, began to emphasize the development of domestic energy resources. Historically, ranchers had grown accustomed to working through the BLM and with energy development companies to accommodate their competing interests. However, the strategic deployment of executive powers, with the support of congressional committees, changed the domestic energy policy landscape of the BLM. In turn, change to BLM's domestic energy policies had the effect of disrupting the historically friendly working paradigm between ranchers and energy developers. The lasting effects of the modern American energy boom are illustrated by the impact split-estate energy development had on the livelihoods of Westerners.

References

Alspach, C. M. (2002, Spring). Surface use by the mineral owner: How much accommodation is required under current oil and gas law? *Oklahoma Law Review, 55,* 89–110.

Anderson, D. (2009). *Split-Estate: What You Don't Know CAN Hurt You.* [Documentary]. United States: Red Rock Pictures.

Associated Press. (2005a, July 11). Split-estate law worried industry. *Casper Star Tribune.* Retrieved from http://www.trib.com/news/state-and-regional/article_22f16910-5778-52ec-a418-626d5da3efef.html

Associated Press. (2005b, December 6). New split-estate law gets test. *Casper Star Tribune.* Retrieved from http://www.trib.com/news/state-and-regional/article_2ed7a174-6c46-5ae7-ae42-1be1061e8c07.html

Bleizeffer, D. (2004a, September 21). Split estates bill takes shape [Web log post]. Retrieved from http://thewesterner.blogspot.com/2004/09/news-roundup-motorcyclists-stop-trail.html#links

Bleizeffer, D. (2004b, October 22). Split-estate parties talk compensation. *Casper Star Tribune.* Retrieved from www.trib.com/news/state-and-regional/article_6882d79b-4d42-52b6-b304-2e0e6fb17ac2.html

Bleizeffer, D. (2004c, November 11). Mineral owners fear surface protection bill. *Casper Star Tribune.* Retrieved from www.trib.com/news/state-and-regional/article_8922dd84-89a7-5da8-94c4-bc43db4898eb.html

Bleizeffer, D. (2005, December 14). State hears first split-estate case; Oil and gas commission chairwoman reprimands lawyers. *Casper Star Tribune.* Retrieved from http://trib.com/news/state-and-regional/article_45771947-5c9f-5b65-96a1-ada48b763243.html

Buckley, F. H. (Ed.). (1999). *The fall and rise of freedom of contract.* Durham, NC: Duke University Press.

Bureau of Land Management. (n.d.). BLM News Room search: Bi-monthly lease sale. Retrieved from www.blm.gov/search/?query=bimonthly+lease+sale&submit=&adv=1&narrow=pr%3Adefault&pr=VALUE_HERE&dropXSL=yes

Bureau of Land Management. (2007). Surface operating standards and guidelines for oil and gas exploration and development. Retrieved from http://www.blm.gov/wo/st/en/prog/energy/oil_and_gas/best_management_practices/gold_book.html

Cater, D. (1964). *Power in Washington.* New York, NY: Random House.

Clifford, H. (2001, November 5). Wyoming's powder keg: Coalbed methane splinters the Powder River Basin. High Country ews. Retrieved from http://www.hcn.org/issues/214/10823

Colorado Surface Owner Protection Act. (2007). LB 1252, Laws of Colorado.

Competitive Leases. (1988). 43 C.F.R. pt. 3120.1-2, 3120.2-2, 3120.3-1, 3120.5-1.

Congressional Hearings. (107th–110th Congresses). Retrieved from http://www.gpoaccess.gov/chearings/browse.html

Davis, C. (2001). *Western public lands and environmental politics* (2nd ed.). Boulder, CO: Westview Press.

Department of Interior: Office of Surface Mining Reclamation and Enforcement. (n.d.). Surface mining law. Retrieved from http://www.osmre.gov/topic/SMCRA/SMCRA.shtm

Douglas, A. R. (1990). *The logic of congressional action.* New Haven, CT: Yale University Press.

Earthworks. (n.d.-a). Landowner stories. Retrieved from http://www.earthworksaction.org/landownerstories.cfm

Earthworks. (n.d.-b). Surface owner protection legislation. Retrieved from http://www.earthworksaction.org/SOPLegislation.cfm

Edelman, M. (1988). *Constructing the political spectacle.* Chicago, IL: University of Chicago Press.

Environmental Working Group. (2004). Who owns the West? Oil and gas leases. Retrieved from http://www.ewg.org/oil_and_gas/

Evans, G. L. (1996). Comment: Texas landowners strike water—Surface estate remediation and legislatively enhanced liability in the oil patch—A proposal for optimum protection of groundwater resources from oil and gas exploration and production in Texas. *Southern Texas Law Review, 37*, 484–485, 515.

Farquhar, B. (2002, November 10). Council told to fight for surface rights. *Casper Star Tribune.* Retrieved from www.trib.com/news/article_24a607ca-7903-516a-8803-28dc85a0992b.html

Fitzgerald, T. (2008). Essays on split estate energy development. Montana State University. Retrieved from http://www.lib.umd.edu/drum/handle/1903/9707

Fitzgerald, T. (2009). *The role of split-estates in coalbed methane production.* (Draft of forthcoming article) Retrieved from http://extranet.isnie.org/uploads/isnie2009/fitzgerald.pdf

Freeman, J. L. (1965). *The political process* (2nd ed.). New York: Random House.

Habermas, J. (1975). *Legitimation crisis.* Boston: Beacon Press.

Hageman, T. S. (1993). Comment: Environmental law: Comparing the effectiveness of oil and gas with coal surface damage statutes in Oklahoma: Bonding producers and operators to land reclamation. *Oklahoma Law Review, 46*, 292.

Hardin, B., & Jehl, D. (2002, December 29). Ranchers bristle as gas wells loom on the range. New York Times. Retrieved from http://www.nytimes.com/2002/12/29national/29METH.html

Hearing on enhancing America's energy security. (2003). *Hearing before U.S. House of Representatives, Committee on Resources, 108*[th] Cong. Retrieved from http://bulk.resource.org/gpo.gov/hearings/108h/85771.pdf

Keffer, W. R. (1994). Drilling for damages: Common law relief in oilfield pollution cases. *Southern Methodist University Law Review, 47*(523), 525.

Kelman, S. (1987). *Making public policy: A hopeful view of American politics.* New York: Basic Books.

Kennedy, J. (2000, July/August). Technology limits environmental impact of drilling. *Drilling Contractor*, pp. 31–35. Retrieved from http://www.iadc.org/dcpi/dc-julaug00/u-doe.pdf

Kingdon, J. (2003). *Agendas, alternatives, and public policies* (2nd ed.). Boston: Longman.

Lands: Interior. Minerals Management: General. (2008). 43 C.F.R. pt. 3000.

Lease of Oil and Gas Lands. (1988). 30 U.S.C. pt. 266. Retrieved from www.law.cornell.edu/uscode/html/uscode30/usc_sec_30_00000226%2D%2D%2D%2D000-notes.html

Levine, M. E., & Forrence, J. L. (1990). Regulatory capture, public interest, and the public agenda: Towards a synthesis. *Journal of Law, Economics, and Organization, 6*, 167–198.

Maas, A. (1949). *Muddy waters.* Cambridge, MA: Harvard University Press.

McCool, D. C. (1987). *Command of the waters.* Berkeley, CA: University of California Press.

McCool, D. C. (1989). Subgovernments and the impact of policy fragmentation and accommodation. *Policy Studies Review, 8*(4), 264–287. https://doi.org/10.1111/j.1541-1338.1988.tb01101.x

McCool, D. C. (1990, Summer). Subgovernments as determinants of political viability. *Political Science Quarterly, 105*(2), 269–293.

McCool, D. C. (1995). *Public policy theories, models, and concepts: An anthology.* Englewood Cliffs, NJ: Prentice Hall.

McCool, D. C. (1998, June). The subsystem family of concepts: A critique and proposal. *Political Science Quarterly, 51*(2), 551–570.

Miller, A. C., Hamburger, T., & Cart, J. (2004, August 25) White House Puts the West on Fast Track for Oil, Gas Drilling. *Los Angeles Times.* Retrieved from www.latimes.com/news/yahoo/la-na-bog25aug25,1,500016.story

Mineral Leasing Act of 1920 as amended (Title 30 of the United States Code § 181 et seq.). Retrieved from www.mrm.mms.gov/Laws_R_D/PubLaws/PDFDocs/MineralLeasingAct1920.pdf

Mitchell, J.G. (2005, July). All fired up: A natural gas boom is transforming public lands in the Rockies, pitting Westerner against Westerner. *National Geographic.* Retrieved from http://ngm.nationalgeographic.com/ngm/0507/feature5/index.html

National Environmental Protection Act of 1969 (Title 42 of the *United States Code* §4332). Retrieved from http://ceq.hss.doe.gov/nepa/regs/nepa/nepaeqia.htm

New Mexico Surface Owner Protection Act. (2007). LB 0827, Laws of New Mexico.

Offe, C. (1985). *Contradictions of the welfare state: Disorganized capitalism.* Cambridge, MA: Cambridge University Press.

Oil and Gas Royalty. (1988). 43 C.F.R. pt. 3103.3-1.

Oil and Gas Leasing, 43 C.F.R pt. 3100-3109 (1988). Retrieved from www.law.cornell.edu/cfr/text/43/part-3100; Lease of oil and gas lands, 30 U.S.C. pt. 266 (1988). Retrieved from www.law.cornell.edu/uscode/html/uscode30/usc_sec_30_00000226----000-notes.html

Oversight hearing on the orderly development of coalbed methane resources from public lands. (2001). *Hearing before U.S. House of Representatives, Subcommittee on Energy and Mineral Resources,* 107th Cong. Retrieved from http://bulk.resource.org/gpo.gov/hearings/107h/75015.pdf

Parks, Forests, and Public Property. (2008). 36 C.F.R. pt. 228, subpt. E.

Ripley, R., & Franklin, G. (1984). *Congress, the bureaucracy, and public policy.* Homewood, IL: Dorsey Press.

Royster, W. (2004, August 30). Stockgrowers dislike split estate draft bill. *Casper Star Tribune.* Retrieved from www.trib.com/news/state-and-regional/article_851a022e-3868-5635-8e5b-1e6b5dca9b67.html

Sievers, L. (2004, July 7). *To drill or not to drill.* ABC News: Nighttline.

Snell, M. B. (2003, July/August). A Cowboy's Lament: Energy companies declare open season on New Mexico. Retrieved from http://www.sierraclub.org/sierra/200307/profile.asp

Stock Raising Homestead Act of 1916a (Title 30 of the *United States Code*).

Stock Raising Homestead Act of 1916b (Title 43 of the *United States Code*). Reservation of Coal and Mineral Rights at § 299. Retrieved from http://assembler.law.cornell.edu/uscode/html/uscode43/usc_sec_43_00000299%2D%2D%2D%2D000-.html

Surface Mining and Control Act of 1976 (Title 30 *United States Code* §§ 1234-1328).

The split-estate law between landowners and oil and gas developers. (2006, October 13). Alexander's gas and oil connections news and trends: North America, 11(19). Retrieved from www.gasandoil.com/goc/news/ntn64101.htm

United States Geological Service. (2000, November). Fact sheet: Water produced with coal-bed methane. Retrieved from http://pubs.usgs.gov/fs/fs-0156-00/fs-0156-00.pdf

United States Environmental Protection Agency. (2000, October). Profile of the Oil and Gas Extraction Industry. EPA Office of Compliance Sector Notebook Project. EPA/310-R-99-006. p.27. Retrieved from http://www.epa.gov/compliance/resources/publications/assistance/sectors/notebooks/oilgas.pdf

Walker, L. M. (1983). Note: Oil and gas: Surface damages, operators, and the oil and gas attorney. *Oklahoma Law Review, 36,* 414.

Western Governors' Association. (2004). Western Governors' Association Coal Bed Methane Best Management Practices Handbook. Retrieved from www.westgov.org/wga/initiatives/coalbed/index.htm

Wilkinson, T. (2005, May 10). Energy boom is crowding ranchers: More ranchers rail against federal 'split estate' laws that control mineral rights beneath their land. *The Christian Science Monitor.* Retrieved from www.csmonitor.com/2005/0510/p01s02-usju.html

Wyoming Surface Owner Accommodation Act. (2005). LB 0070, Laws of Wyoming.

Drilling Pad with Retention Pond on Privately Owned Farmland

The Expansion of Hydraulic Fracturing in the West Was Often Marked by Independent Wildcat Operations

Drilling Rigs In Close Approximation to the Wind River Wilderness Range and the Town of Pinedale in Wyoming

The Impact of the Expansion of Hydraulic Fracturing is Exemplified by Rig-to-Rig Spacing Proximity at the Ground-Level

Cattle on the Open Range Standing in Unlined Disposal Pond for Produced Fluids from Fracking Operations; A Common Practice at the Start of the Boom

The Confluence of Ranching and Hydraulic Fracturing Operations on Privately Owned Ranching Property

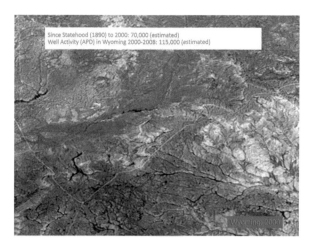

Wyoming Landscape Pre-Expansion of Hydraulic Fracturing in the West From an Aerial Point-of-View (2000)

Wyoming Landscape Post-Expansion of Hydraulic Fracturing in the West From an Aerial Point-of-View (2007)

Colorado Landscape
Pre-Expansion of
Hydraulic Fracturing in the
West From an Aerial
Point-of-View (2000)

Well Activity (APD) in Colorado 2000-2008: 73,000 (estimated)

Colorado Landscape
Post-Expansion of
Hydraulic Fracturing in the
West From an Aerial
Point-of-View (2007)

A Different Perspective of
Colorado Landscape
Post-Expansion of
Hydraulic Fracturing in the
West From an Aerial
Point-of-View (2007)

Similar to Wyoming and
Colorado, New Mexico
Landscape Post-Expansion
of Hydraulic Fracturing in
the West From an Aerial
Point-of-View (2007)

A Different Perspective of
New Mexico Landscape
Post-Expansion of
Hydraulic Fracturing in the
West From an Aerial
Point-of-View (2007)

A Different Perspective of
New Mexico Landscape
Post-Expansion of
Hydraulic Fracturing in the
West From an Aerial
Point-of-View (2007)

The Proximity of Hydraulic Fracturing Operations to the Town of Pinedale, Wyoming Against the Backdrop of The Wind River Wilderness Range, Home to Wyoming's Highest Peak (Gannett, 13,810 ft) and The Largest Glacial Formation in the Continental U.S. and Headwaters to The Three Major U.S. River Drainage Systems: (1) Wind-Missouri-Mississippi, (2) Snake-Columbia, and (3) Green-Colorado

Chapter 5
Governance

Abstract The disruption of a subgovernment results from numerous conditions. From the perspective of federal political appointees to the Department of the Interior (DOI) and career administrative officials within the Bureau of Land Management (BLM), disruption of the BLM's land-use subgovernment resulted from a variety of conditions. In the opinion of these government officials, energy development spilled onto the privately owned surface lands of split-estate property across the Western United States as the Bush Administration directed expansion of domestic energy development. In turn, the leasing and permit granting of split-estate properties for energy development disrupted the stability of the existing BLM land-use subgovernment. Among the government officials interviewed, there is general consensus that the disruption generated competition between ranchers and energy developers for control of the BLM's land-use subgovernment. To these government officials, the end result of the conflict and competition is clear: Energy interests have displaced ranching's dominance of the BLM's land-use subgovernment.

Keywords Political appointees · Bureau of Land Management · Department of Interior · Bureaucracy · Split-estate energy development · Political conflict

> "...yeah, it's a shame that the guy [energy developer] gets to come on your land, and it seems like a shame that you [surface owner] have to let him, and it seems like a shame he is really the predominant user, but you know, he really doesn't owe you anything other than to reclaim your landscape on the surface to what it was...." (Don Simpson, Director, State of Wyoming BLM)

The Voice of Government

The disruption of a subgovernment results from numerous conditions. From the perspective of federal political appointees to the Department of the Interior (DOI), career administrative officials within the Bureau of Land Management (BLM), and

© Springer Nature Switzerland AG 2019
R. E. Forbis Jr., *Altered Policy Landscapes*,
https://doi.org/10.1007/978-3-030-04774-0_5

at least one elected state representative,[1] disruption of the BLM's land-use subgovernment resulted from a variety of conditions. In the opinion of these government officials, energy development spilled onto the privately owned surface lands of split-estate property across the Western United States as the Bush Administration directed expansion of domestic energy development. In turn, the leasing and permit granting of split-estate properties for energy development disrupted the stability of the existing BLM land-use subgovernment. Disruption of the BLM's subgovernment effectively triggered a political conflict between ranching and energy interests as each interest sought to protect its use of lands and resources. Among the government officials interviewed for this research, there is general consensus that the political conflict generated competition between ranchers and energy developers for control of the BLM's land-use subgovernment. To these government officials, the end result of the conflict and competition is clear: energy interests have displaced ranching's dominance of the BLM's land-use subgovernment.

Motivated by the executive and legislative decision-making referred to in previous chapters, the BLM shifted its land management policies to emphasize domestic energy development. Among the government actors interviewed here, each remains deeply immersed in the evolving controversy of split-estate energy development in the Rocky Mountain West. In the opinion of these governmental actors, the problems associated with split-estate energy development remain complex, dynamic, and troubling. The voices represented here are those of governmental elites; they are a select sample of actors who interacted within the networks of the BLM's land-use subgovernment at the highest levels during the period of time in which domestic energy development expanded across the Western United States.

---

[1] Note: (1) Bureau of Land Management (BLM) participants include Pat Shea, former Director of the BLM under President William J. Clinton; Don Simpson, State Director of Wyoming BLM; Larry Claypool, Deputy State Director of Minerals and Lands Wyoming BLM; Lynn Rust, Deputy State Director of Minerals and Lands Colorado BLM; and Tony Herrell, Deputy State Director of Minerals and Lands New Mexico BLM.

(2) Department of Interior (DOI) participants include Rebecca Watson, former Assistant Interior Secretary for Lands and Mineral Management under former President George W. Bush, and an unnamed DOI political appointee under former President George W. Bush.

(3) State Representative participation was, unfortunately, limited to one interview, that of Colorado State Representative Ellen Roberts (R-Dist.59). Numerous attempts were made and strategies employed to gain access and interview with legislative sponsors of surface owner protection acts in both the states of Wyoming and New Mexico. In either case, e-mails and telephone calls went unanswered. One reason for this might be that unlike Colorado, Wyoming and New Mexico legislators are part time. Access to Rep. Roberts was gained by just showing up in her offices at the state capital and working with her assistant to reserve a time for the interview. Even then, the time Rep. Roberts could spare was very limited, and, as a result, the interview is the shortest among all the interviews conducted. Nonetheless, her viewpoint is reflective of journalistic accounts of the political battles and lobbying efforts undertaken by ranching and energy interests in the states of New Mexico and Wyoming.

Energy as a Policy Objective

The Bush Administration's policy objective of expanding domestic energy development was the result of an energy resource shortage, technological advancements, and market costs associated with limited energy supplies. According to Rebecca Watson, former Deputy Secretary of the Department of Interior for Lands and Minerals under former President George W. Bush:

> In 2002, 3, and 4 there was a natural gas shortage, I'm a firm believer in the market, and it was demonstrated that you had Chairman Greenspan testifying to Congress in 2003 about the impacts on the economy of natural gas shortage. Obviously, a shortage in our market economy drives up the price of natural gas. And so, natural gas was in short supply. There was a need to get it. That was something the Bush Administration was quite focused on 'cause we were seeing the loss of chemical industries were going overseas, fertilizer industries, it was having a huge impact on the agricultural economy because all of those are heavily dependent on natural gas. Ethanol, heavily dependent on natural gas. So there was a drive, an important social goal to get more natural gas into the system. (R. Watson, personal communication, June 16, 2009)

According to Ms. Watson, pursuit of energy resources increased, and energy costs were reduced by the advent of fracking and directional drilling. The difference, according to Watson, between conventional and unconventional energy resource development allowed for the administration to intensify domestic energy development in areas where energy development had once been considered impossible. As Watson notes:

> The other thing that supported [development] was the Department of Energy had done research in the late 80s and early 90s on how you can release this unconventional natural gas. And unconventional natural gas was in tight sands, in coal, the coal-bed methane or coal-bed natural gas. And, that was something that was a huge resource, but it was not able to be developed. That began to be developed, fracking is fundamental to that, the ability of fracturing this tight rock to release the gas. And directional drilling really didn't come until I would say 2004 or 5, that's when they were really able to maximize the use of directional drilling. (R. Watson, personal communication, June 16, 2009)

Prior to these technological advancements, development on the surface was intense as the energy industry sought to take advantage of market prices for energy. In essence, as government and industry responded to resource shortages in energy markets, the pace of energy development was permitted to speed up prior to perfecting the technology of directional drilling. As Watson observed, the combined effect of energy markets and technological advancements led to energy development "on quite tight spacing, lots, lots of straws to get the gas…and that intense surface development that was different than what ranchers and surface owners were used to" (R. Watson, personal communication, June 16, 2009).

From Watson's perspective, it was the intensity of capturing energy resources from unconventional areas led to the issue of split-estates. According to Watson, "The thing that struck me the most after I left Interior was the difference between unconventional natural gas development and conventional natural gas. And I think that contributes to the issues of split-estate. Not so much the fact that estates are split, because people have been dealing with that for a long time, but the fact that in order to develop unconventional natural gas it's much more intensive on the surface" (R. Watson, personal communication, June 16, 2009). But, energy companies cannot drill without an Approved Permit to Drill (APD), and, while energy development activity was on par with the Clinton Administration, Ms. Watson notes that "what changed was the development; the actual issuance of permits to drill" (R. Watson, personal communication, June 16, 2009). According to Watson, "there were more permits to drill by quite a few, but again that comes out of the fact that we're dealing with unconventional natural gas. We, you have to have a number of permits. For each well you need a permit. You don't have a permit for multiple wells until you get that directional drilling phase" (R. Watson, personal communication, June 16, 2009).

In the early years of the Bush Administration, rising energy shortages and costs were addressed by efforts to expand domestic energy development. This meant that APDs had to be issued quickly. And this meant directing the BLM to expedite the APD administrative process. In Watson's opinion, if the Bush Administration was going to effectively increase energy resources and lessen energy costs, the Administration would have to increase the number of APDs being approved by the BLM. According to Watson, "That's why expediting energy permits was put in there. Because even with the so-called expediting, which if you look at, it never got that expedited. It could never meet; it could never match the demand for permits that was there in the industry. The industry wanted permits" (R. Watson, personal communication, June 16, 2009). Legally, the BLM is required to adhere to the National Environmental Policy Act's (NEPA) environmental and collaborative planning process. And as Watson notes, that process takes substantive amounts of time with "no clock on it" and "goes for as long as it's needed" (R. Watson, personal communication, June 16, 2009). And, according to Watson, "Industry never understood that. They wanted a tight clock and even though permitting accelerated it never matched that demand. And, I don't think I fully appreciated [until some years into it] how unconventional gas changed things, the pace of development because of the permits you needed" (R. Watson, personal communication, June 16, 2009).

The federal expansion of domestic energy development was driven by three factors: energy markets, technological advancements, and political willpower. No amount of political will, rising energy costs, or advancements in drilling technology could have prepared government officials for the political conflict that emerged between ranching and energy developers. According to a senior Department of Interior appointee during the Bush Administration, who requested anonymity, this was particularly true of the BLM as it responded to President Bush's executive orders. As this political appointee notes, "There's no question, they [BLM] were under tremendous pressure to get these APDs issued" (Unnamed DOI political appointee, personal communication, May 26, 2009). According to Rebecca Watson, the idea that the BLM would respond as desired by the Administration regarding the APD approval process is "unrealistic" (R. Watson, personal communication, June 16, 2009). In Watson's opinion, "the idea that the President writes an executive order and everyone snaps to and charges off, that's just unrealistic. But, yes energy was made a priority because the President and Cheney thought it was a priority for our economy and our economic well-being. So that was important and that message was clearly transmitted to people, that energy development was a critical issue" (R. Watson, personal communication, June 16, 2009).

The scope and pace of domestic energy development would eventually lead to increased leasing and development of split-estate lands. According to the unnamed DOI appointee, given the historical nature of the ranching and energy alliance, the fact that conflict arose between these two groups over split-estate energy development was surprising. According to this DOI official:

> I remember early on in 2001 that there were some landowners who were very concerned about the conflicts. I don't think anybody was aware that there was going to be these kinds of conflicts and, too, I thought they [energy industry] would be good neighbors. From my perspective why would you go out and antagonize the ranching community which historically have been, conservative Republicans, and the oil and gas industry which has historically been conservative Republicans. Why would they [ranching and energy] go politically head-to-head and create the kind of conflict when they could resolve themselves by sitting down and working together? (Unnamed DOI political appointee, personal communication, May 26, 2009)

Creating conflict between ranchers and energy developers was an unintended consequence of the Bush Administration's effort to expand domestic energy development. The origins of the conflict, as was discussed in Chap. 2, can be traced back to the enactment of federal homesteading laws as well as federal minerals and grazing laws. As the senior DOI official notes, "this fundamental conflict traces back to the 1916 [Stock-Raising] Homestead Act and to the decision as to who retained the oil and gas" (Unnamed DOI political appointee, personal communication, May 26, 2009). Created by the Stock-Raising Homestead Act of 1916, the 58 million acres of Western split-estate properties are emblematic of a decision made by the federal government to retain the oil and gas. The decision to retain the oil and gas underlying the surface estates of homesteaders was well-intentioned for its time. Good intentions notwithstanding the creation of split-estate properties and the federal government's retention of the oil and gas within the subsurface estates of those properties triggered conflict between ranchers and energy developers in the twenty-first century.

The Cozy Relationship of Ranching and Energy Development

The legislative origins of split-estates are the result of federal legislation intended to secure mineral resources for the benefit of the public. According to Pat Shea, former Director of the Bureau of Land Management (BLM) under President William J. Clinton, the creation of ranching and energy development interests developed concurrently (P. Shea, personal communication, June 2, 2009). The origin of ranching and energy's harmonious relationship merged into what is commonly referred to as the "strong corner" within the BLM's land-use subgovernment. The agreeable nature of this alliance is reflective of each group's vested economic interests in the public domain. The interest of ranching and energy as well as their ability to control federal policy decisions is also a reflection of federal efforts to regulate their dominant but shared use of the public domain. As Pat Shea notes:

> It really strikes me that BLM reflects its (merging of ranching and energy development) origin. It really began in the nineteenth century as the General Land Office and therefore, as they grew, [they] got dumped, so to speak for administrative purposes into the General Land Office. And then in 1948 when BLM was created there was an effort to consolidate. And then with FLPMA (Federal Lands Policy and Management Act of 1976) there was this sense that somehow you could put them together and in many areas that worked out quite well. (P. Shea, personal communication, June 2, 2009)

Shea also notes that the cozy relationship that emerged between ranching and energy developers was a conflict waiting to happen as "the tension between the different constituencies who have an economic dependency on BLM and its policies, ha[d] not been worked out and the split-estate is legal, both by judicial decision and by statutory legislation [and is now] a great example of that split" (P. Shea, personal communication, June 2, 2009). The reality of split-estates is that between the competing economic interests of ranching's surface use and energy developer's subsurface use, each interest desires favorable political, legal, and administrative decision-making.

As was discussed in Chap. 2, from the perspective of history, governmental decisions regarding the economic interests of ranching and energy developers corresponded as laws, legal decisions, and regulations were developed. During the period of time in which land management laws were created, the economy of ranching and energy was roughly equivalent, as was their respective use of the public domain, and conflict was avoided. Conflict between ranchers and energy developers did not occur until federal efforts to expand domestic energy development encroached onto private lands where the economic interests of the two interests come into direct conflict with each other. The conflict is perpetuated because of the vested economic and property interests that are shared between property-owning ranchers, energy developers, and the federal government. In many instances, the types of conflicts that emerge between these parties are the result of not having resolved the existing political, legal, and administrative tensions inherent to split-estate energy development.

The contemporary conflict and competition between ranching and energy development are based, therefore, upon unresolved legal, legislative, and policy questions concerning split-estate energy development. These questions remain unresolved because, from the perspective of administrative decision-making, regulation of split-estate energy development is dependent on legal and administrative interpretations of the late nineteenth and early twentieth-century land management legislation. Thus, the modern administrative reality of regulating split-estate energy development is that, "in most split-estates of competing or conflicting interests, nobody at the end of the day is going to be happy. The winner, if you will, is not going to be happy 'cause they didn't get everything they asked for, and the loser is just going to say you made the wrong decision" (P. Shea, personal communication,

June 2, 2009). Additionally, more unhappiness results from increased numbers of split-estate property owners throughout the Rocky Mountain West. From the period of homesteading to the present time, most large tracts of private ranchlands have changed hands numerous times. As ranchlands were subdivided, mineral rights were often sold or retained independent of surface properties.

As a result of multiple exchanges of property, there is little clarity regarding the decision-making realities of owning private property in the Western United States. The reality is that decision-making power regarding use of property, especially where the surface and mineral estates have been split, is shared among multiple stakeholders. Once the administrative decision has been made to develop energy on the split-estate and the energy lease to develop the federally owned mineral estate has been sold at auction, decision-making regarding the use of the privately owned surface is a matter of negotiation between the surface owner and the energy developer. Here again, the late nineteenth- and early twentieth-century land management legislation dictates that development of the mineral estate is dominant over development of the surface estate. Federal and state courts have consistently upheld the mineral estate's dominance. As a result, the BLM has promulgated rules and regulations that effectively shield energy developers' vested property interest in federal energy leases. The BLM enforces the legal dominance of the mineral estate through administrative rules and regulations designed to protect the energy lease as a property interest. The means by which the BLM protects the lease is by allowing energy developers to post bond if an agreement for access and use of the surface estate cannot be successfully reached with the landowner. In the case of split-estates then, mineral estates and energy leases take precedent over the private use and enjoyment of the surface estate. Thus, while there are multiple economic and property interests at stake in the development of a split-estate energy lease, the dominance of the mineral estate is politically, legally, and administratively protected.

Property Rights: Decision-Making

Decision-making regarding one's private property has never been absolute. But, among the property-owning public, there is the reasonable expectation that decisions regarding property access and development are theirs alone to make. When split-estate energy development occurs, preconceived notions of property rights lead to the stunning realization among the West's property owners that shared ownership is, according to Don Simpson, Director of the State of Wyoming BLM, "an accident [of] history" (D. Simpson, personal communication, March 23, 2009). As Simpson notes:

> When Farmer Jones or Homesteader Bob got their property and they got their 160 or their 320 or their 640 (acres), there was some reservation of in there to the United States for minerals. So now, all of a sudden, your great-granddad sells to his father, and his father sells to my friend, and then I buy half of it from him, well, I'm busy enjoying my 40 acres now instead of the 640. I don't know your granddad that initially bought the property, and I think

it's safe to say that 9 out of 10 people who get that property, if there's no mineral
development occurring around it, probably have no clue that they're not picking up the
mineral rights. I'll bet if you go ask a hundred people downtown or at the grocery store do
they own the minerals or not, they'd probably go, "I don't know. I have no clue."
(D. Simpson, personal communication, March 23, 2009)

Property owner confusion is compounded by what many BLM administrators call
the "urban interface" (L. Rust, personal communication, May 19, 2009). According
to US census statistics, since the 1970s, the states which compose the Rocky
Mountain Region of the Western United States have experienced almost unfettered
population growth.[2] The size of urban and rural population centers across the Rocky
Mountain West expanded as more people relocated to the Rocky Mountain West.
One variable that significantly contributed to the growth of the urban interface was
the demise of large family-owned ranching and farming lands. As generations of
Americans left the life of ranching and farming, formally large unified tracts of
lands and resources were subdivided and sold. The newly subdivided lands were
then purchased by persons relocating to the states of the West. Over time, this subdivision of land has had the effect of confusing which estate is, and which is not,
controlled by individual property owners.

Homesteading laws were designed to encourage population growth in the West.
Federal officials could not have anticipated how the West would eventually be settled. Larry Claypool, Deputy State Director of Minerals and Lands for Wyoming
BLM, comments that "It's interesting because you look back through history and I
don't think the Stock Raising Homestead Act really foresaw the subdivisions and
what happened in the future. You start dividing all that stuff up and selling to each
and every person and you know, you see what's on the surface, but what's underneath it is lost. It just doesn't carry that same weight in the historical times as it does
now. It just wasn't as important" (L. Claypool, personal communication, March 23,
2009). The subsurface mineral estate of Western ranchlands was unimportant at the
time because energy sources were located elsewhere, relatively plentiful, and more
easily developed. Economically and technologically, Western energy resources
developed during the early twentieth century did not overtly intrude upon public or
private lands suitable for grazing livestock or growing crops.

During the early twentieth century, cattle grazing dominated the economy of the
West. It is during this period of time that grazing leases were economically beneficial to the federal government as a source of revenue. Simply put, federal revenue
generated from grazing leases outperformed revenue generated from the sale and
development of energy leases during this period of time. The economics of cattle
and energy, however, only partially help explain why energy development did not,
until the present time, conflict with ranching interests. The technological challenges
associated with developing energy resources of the time must also be considered as
an additional explanation for why ranchers and energy developers did not come into
conflict for such an extended period of time. Technologically, developing
easy-to-reach domestic energy resources was economically advantageous to the
bottom lines of the oil and gas industry. This remains true today.

---

[2] Note: US Census Bureau, 1970–2009; population census figures and estimations of New Mexico,
Colorado, and Wyoming (1990–2010) retrieved from www.census.gov.

Domestic energy resources that were once easy to access and develop are now played out. The domestic energy resources that remain are sources that have remained relatively off-limits in terms of the economy and technology of developing untraditional energy resources such as coalbed methane (CBM). Because of limited energy resources, developing these non-traditional energy resources is now economically advantageous to the energy industry. Technological advancements have been made, and new, previously undeveloped sources of energy are now available to the energy industry, and energy developers sought to take advantage of favorable market conditions, and new technologies came online at roughly the same time. This turn of events allowed drillers to develop non-traditional energy resources in previously inaccessible places.

The economics and technological advancements of the twenty-first-century energy development coincided with political willpower favoring expanded domestic energy development. In essence, the modern energy boom of the late twenttieth and early twenty-first century created a "perfect storm."[3] The sale of grazing leases as a means of generating federal revenue could no longer sustain itself in the face of the revenue generated by energy development. As more and more non-traditional sources of energy were opened to development, federal revenue generated by the sale of energy leases—as well as the federal royalties derived from their development—far outpaced revenue generated by grazing leases. From the perspective of economics, the energy boom that had begun in the late 1990s had overtaken grazing as a means of generating revenue by the start of the Bush Administration. And, as more areas were opened to energy development, that development spilled over and onto the split-estate lands of Old West ranchers and New West homeowners.

Energy and Urban Development

Domestic energy development is cyclical. Energy development in the West is as infamous for its episodic energy booms as it is for the certainty in its eventual energy busts. The boom and bust of the West's energy cycles played a significant role in property owner confusion over who owned what as twenty-first-century domestic energy development expanded across the West. As is noted by Don Simpson, "If there's no mineral development occurring around it, [property buyers] probably have no clue that they're not picking up the mineral rights" (D. Simpson, personal communication, March 23, 2009). The West's previous energy boom of the 1970s did not see similar conflicts emerge between surface owners—primarily ranchers—and energy developers. One explanation for low ratios of conflict is that during the 1980s and the better part of the 1990s, domestic energy development had waned as

---

[3] Note: Numerous governmental and nongovernmental participants used the phrase "the perfect storm" as they responded to questions regarding the confluence of the energy market, technological advancements in energy development, and Bush Administration activities aimed at expanding domestic energy development.

energy markets slumped. Coincidently, this was also the period of time when populations across the West experienced their most significant growth. Thus, an extended period of low rates of energy development coincided with an extended period of population growth and urban development. This pattern of low rates of energy development and high rates of population growth continued throughout the first decade of the twenty-first century. According to Lynn Rust, Deputy State Director of Minerals and Lands for Colorado BLM:

> I've been in this business since 1977, so over 32 years of now of regulation of federal oil and gas and other minerals also. I've seen it [Western energy development] swing back and forth. In the late '70s oil and gas activity was very high, then you had the crash that started in the early'80s, continued through the mid '80s for sure, just crashed bad, its economy, people flooded out because of drop in prices, and the industry kind of went into the hole, and then it started coming back, and so it is a very cyclical industry. (L. Rust, personal communication, May 19, 2009)

Thus, when contemporary population growth occurred, a new pattern of land ownership interfaced with the historic pattern of energy development. The urban interface referred to by BLM administrators and the rise in land-use conflict is a result of this changed pattern of growth and development. Lynn Rust comments that "What has changed in a lot of aspects, it used to be we, the BLM, were dealing primarily with ranchers and the ranchers didn't have as big of concerns, but as the West is growing more populated, the urban interface issue is really growing, and more and more the split-estate issue involves private surface owners that aren't ranchers. Out there with what I call ranchettes, maybe 10, 20, 30, 40 acres, their little paradise. Maybe they earn their living as an Internet cowboy right from their home" (L. Rust, personal communication, May 19, 2009). While the face of Western landownership has changed over time, homesteading legislation reserving the right to develop the mineral estate has not.

The federal mineral estate remained legally dominant during the period of time when the number of split-estate property owners increased. And because the mineral estate remained dominant, the administrative rules and regulations that guide the process of split-estate energy development remain unchanged as well. That split-estate energy development has not been reformed by the federal government should not come as a surprise. This is because split-estate property ownership increased during a bust cycle in domestic energy development. Simply stated, limited domestic energy development meant little if any conflict with split-estate property owners. Conflict between surface owners and energy developers does not occur until the boom cycle of domestic energy development returns in the late 1990s.

As the twenty-first-century boom cycle of domestic energy development took hold, conflict with split-estate property owners increased. Conflict between split-estate property owners and energy developers increased as the number of wells being drilled on ranchlands increased. This time, however, energy developers were not just dealing with Old West ranchers; they were also dealing with New West ranchers. As domestic energy development expanded under the Bush Administration, the antiquated nature of split-estate energy development was ill-equipped to avoid triggering conflict between ranchers and energy developers.

Administrative Procedure for the Development of Energy

Today when parcels of land are nominated for sale by energy interests, a detailed process of land management planning is begun by the BLM. Once the land management plan has been approved, the auction and sale of the energy leases take place. Prior to the lease sale, the BLM is required to give a 45-day public notice of the impending lease sale in order for any protests to be weighed by the BLM's field office. The cutoff for filing a protest is 15 days prior to the lease sale. However, BLM notification of individual landowners affected by the lease sale is not required. This means that owners of split-estate properties where energy development has been proposed are not contacted directly by the BLM prior to the auction and purchase of energy lease(s).

According to Lynn Rust, "We [BLM] publish it in the Federal Register. We post the list in our public room(s). We put out press releases. We mail individual booklets to anybody. They cost five bucks. Who wants one? We also send a letter to each county commission that has parcels for sale, notifying them. We also notify each Oil and Gas Conservation Commission liaison in each county and they post them to their website. So, we [Colorado] go quite a bit beyond what we're required to do as far as trying to get notification out there" (L. Rust, personal communication, May 19, 2009). Consequently, split-estate landowners must be attentive to BLM public notifications of impending lease sales should they wish to file a protest. Again, according to Lynn Rust, "As far as the split-estate issue, there is still concern. Some of the property owners are calling for [is that] they want to be individually notified that parcels under their property are going to be put up for sale. But at this point, that's not required" (L. Rust, personal communication, May 19, 2009). According to Rust, "[That] would be a difficulty on us because we don't track private property transactions. In other words, if somebody owned 640 acres out there and decide to subdivide it to 40-acre ranchettes, we don't know that: We only know there's still one guy who owns 640 acres" (L. Rust, personal communication, May 19, 2009).

Unless new property owners throughout the Rocky Mountain West understand the history of their property's ownership, and are attentive to any potential energy lease sales occurring in their area, they would likely be unaware of their subsurface estate being sold at auction for the purpose of developing energy resources. These property owners remain unaware of their property's potential for energy development until, as required by law, a representative of the company, commonly referred to as a "landman," contacts the property's owner by certified letter. Once contacted, property owners have 45 days to respond to the company's notice of intended exploration and development. At the same time, the company seeks the required Approved Permit to Drill (APD) from the BLM if the mineral estate is federally owned. Once the APD has been approved, an onsite pre-drill inspection occurs where the surface owner is invited to attend either by the BLM or the developer.

It is during the pre-drill inspection that a surface owner can express any concerns to the BLM administrator regarding the proposed development activity (L. Rust, personal communication, May 19, 2009). In addition to pre-drill inspection, the

BLM also requires that either a signed Surface Owner Agreement or a certification that there is an agreement in place be filed with the agency. The BLM "encourages the industry [to] get a Surface Owner Agreement worked out [because] we don't want to have to go to the bond on process" (L. Rust, personal communication, May 19, 2009). As was discussed in Chap. 4, if a Surface Owner Agreement cannot be reached, energy developers can simply post a bond to access the privately owned surface estate.

The consensus opinion among BLM administrators is that "bonding on," as the process is commonly referred to among those familiar with the process, is rare. And while DOI appointees and BLM administrators expressed concern over how surface owners were being treated by energy developers, BLM administrators were adamant in expressing that they had "no role" in the negotiation process (R. Watson, personal communication, June 16, 2009). Thus, when it comes to negotiations over Surface Use Agreements, BLM administrators do not engage outside the legal boundaries of mandated legal oversight because "The regs follow the law, fair or not. If somebody needs to change it, the law needs to be modified" (D. Simpson, personal communication, March 23, 2009). Until the law is reformed, BLM administrators will continue to be "good soldiers" as they respond to the "political agendas of Congress and of whatever administration, whoever is in the White House. [Because] each administration looks at things in their own way" (L. Rust, personal communication, May 19, 2009). Or, as Rebecca Watson quipped, "Yeah, they [BLM administrators] don't want to be involved in blessing or cursing people's ills" (R. Watson, personal communication, June 16, 2009).

Unregulated Surface Owner Agreements

The conflict between surface owning ranchers and energy developers is most apparent during the period of negotiation to secure a Surface Owner Agreement. In the past, ranchers and developers had amicably resolved differences, creating the conditions for energy development to occur. Because the BLM had been politically directed to emphasize domestic energy development during a period of rising energy market prices and technological advancements in energy resource extraction, developers urgently sought to take advantage of the favorable conditions. In doing so, some companies and their representatives were overly aggressive in their approach to negotiating Surface Owner Agreements with split-estate property owners. Often companies sent landmen from other states such as Oklahoma and Texas to Western states to negotiate the terms of the agreement. Energy development in Oklahoma and Texas differs from the West in that split-estate federal ownership is a rarity, so these representatives were unfamiliar with federal requirements as well as Western traditions regarding energy development on privately owned ranch lands.

Unlike previous time periods when the traditional alliance of ranching and energy development worked well, there is now a cultural difference that divides ranchers and energy developers. According to Tony Herrell, Deputy State Director of Minerals and Lands for New Mexico BLM, "There's a cultural difference" (T. Herrell, personal communication, May 20, 2009). From Herrell's perspective of oil and gas developers, "It's fast-paced; they have so many things to put in place. It's a very complicated process. Between the environmental permitting government agencies like ours, and the bureaucracy that goes with it, it's very frustrating for them. And so they feel kind of blocked from it [energy resource], so it does feel like a real battle from their perspective" (T. Herrell, personal communication, May 20, 2009).

From Herrell's perspective, energy developers are more likely to be concerned with "permits, lawsuits, and not being able to develop here, or actually being in court" (T. Herrell, personal communication, May 20, 2009). As Herrell has observed, when negotiations begin "[Industry is] talking usually about money and costs, the survivability of the company." Ranchers, on the other hand, have different concerns, "[They] usually talk about family; they talk about your family; they ask how you are doing, and then eventually they'll get down to business" (T. Herrell, personal communication, May 20, 2009). In Herrell's opinion, "When you sit down at the table with different groups of folks, you can see the train wreck that can happen. They're [industry] feeling desperate; they need to get some cash flow, and if they have a known resource that they know they can make some money off, they're wanting to get in there and get the deals signed, and the rancher wants to think about it for a while, and so you can see the train wreck. You can feel it when you're around them" (T. Herrell, personal communication, May 20, 2009).

Most split-estate surface landowners are treading across unfamiliar legal and regulatory territory. The less experienced the ranch owner is, the less likely he or she is to be intimately acquainted with the legality and regulatory requirement of energy extraction. As Rebecca Watson notes, "[Ranchers] are sophisticated. They understand their land, what mineral rights they have and don't have. These ranchette buyers aren't sophisticated" (R. Watson, personal communication, June 16, 2009). Unfamiliarity with the law is problematic for split-estate property owners because compensation for nuisance issues or economic losses associated with energy development activities is not regulated with any specificity.

Specific compensation remains unregulated because, as was argued in Chap. 2, Surface Owner Agreements are composed from the liberty-of-contract legal paradigm where the property interests of the parties are theirs to protect through contractual negotiation and agreement. For example, in the opinion of Rebecca Watson:

That's the big question. The rule of law is what our country was founded on. The other thing our country was founded on was property rights…that ability to own a piece of land gave [people] power and liberty as a citizen. They could make their fortune. They had a voice in government. And those are very fundamental, important, concepts. If we lose those then anything can happen and so I think that, yes, you enter into contracts with people, but I think that anyone would agree in a contract situation that you need information; you need understanding of the laws in order to do that properly. But I think on both sides of it, both sides again have property rights. Those property rights are protected via a contract and you should be bargaining together, have the information you need and come to an agreement that works for both sides of it. And then that agreement should be respected. I think it's fundamental for our country and what makes us different than a despotic regime in other places where government can decide to take property or to destroy contractual relationships. (R. Watson, personal communication, June 16, 2009)

Defense of one's interest is dependent on familiarity with the law, and most split-estate landowners, when first approached by the energy developer, lack any understanding of the laws or regulations that guide split-estate energy development. The advantage then goes to the party who best understands the law, and, as is most often the case, the advantage in negotiating Surface Use Agreements is with the energy developer's representative: the landman.

Landmen and Split-Estate Property Owners

Unfamiliarity with Western social traditions among landmen from other parts of the country created animosity with ranchers. As noted by Rebecca Watson, "I think that in the rush to develop natural gas, the bringing in of some of these people that were not familiar with the culture of the West from Texas and Oklahoma, people's toes were stepped on, people were treated poorly and people talk in small western states and that created some problems" (R. Watson, personal communication, June 16, 2009). While energy developers were at times their own worst enemies when it comes to approaching split-estate landowners, Watson still believes that "By and large the oil and gas industry tries to have a very good relationship with ranchers." Given the pace and scope of energy development occurring on split-estates, the historical relationship between ranching and energy is at risk if landmen, representing the interests of the energy company, do not act in a responsible manner when negotiating Surface Use Agreements with property owners. According to Watson, "Most companies, responsible companies, understand and work hard at [maintaining the relationship]" (R. Watson, personal communication, June 16, 2009). BLM administrators are quick to note that erosion of the traditional notion of who is a rancher and who is not a rancher has made it difficult for energy developers to maintain the friendly nature of their relationship with ranching communities.

If the pace of domestic energy development continues to steadily increase, energy developers are going to have to account for how their representatives interact with landowners. As Lynn Rust notes:

> First of all, there's always room for improvement, and again, your smart companies, they figure it out pretty quickly, the best way to be operating. Obviously you're going to interact with an old rancher who has been out there, you know, since Christ was a corporal as you are with a fairly new subdivision of 20-acre ranchettes of basically new urbanites who wanted to escape the city and live out in their paradise. If I were a major corporation, say wherever, based out of Houston, I would never send people out of Houston to go talk to these people, never…As far as dealing with the United States, it would be don't lie to us, which, you know, I've had that done to me. 'Oh, yeah, we've got this [Surface Use Agreement]; no problem.' Then you find out that's not the case. I've been met with a gun before because the operator said, 'oh yeah, we've got a Surface Owner Agreement, yeah, yeah.' We [BLM] go out there and it's like 'who the hell are you? Wait a minute.' So…. (L. Rust, personal communication, May 19, 2009)

In the opinion of others, the animosity cuts both ways. Often, because of the unregulated nature of Surface Owner Agreements, a relative minority of split-estate property owners will attempt to take advantage of a developer's desperation to develop the energy resource prior to a drop in market price. For example, according to Don Simpson, other than the "5 people [that] squawk [among] the other 5,000 [that] don't" or the "5 percent, or whatever, that were kind of thinking I want more than that saying 'My God, this company comes on my land and they're making millions of dollars, and I get, you know, they replace my gate and build a pond for me.' So I think there was a vocal minority" (D. Simpson, personal communication, March 23, 2009).

To other BLM administrators and at least one state elected official, the term "minority" is a relative term. In some instances, animosity and conflict are relative to the scope and pace of the energy's development within a particular area. These areas are where the nature of the energy's development is in direct conflict with ranching interests and, as a result, where animosity is most palpable. For example, in Colorado, where the urban interface is greatest, the level of conflict depends on where the energy development is taking place. Colorado State Representative Ellen Roberts (R-District 59) noted that, "La Plata County is the largest producer of natural gas in Colorado [and] most of the production has been on private lands" (E. Roberts, personal communication, March 25, 2009). In her experience, conflict between ranchers and energy developers was not limited to a relative minority of constituents. When asked whether she agreed with the assessment that conflict was limited to just a small, rather vocal, group of dissatisfied persons, Rep. Roberts responded, "No, no, I don't agree. Whether it's the noise, the dust, the trucks on the road, whatever, there will be a day when they leave, and my concern was the public health and the historic background that the mineral owner had the dominant power, and things that people were willing to sacrifice 100 years ago, I don't think people who live here today are willing to sacrifice that" (E. Roberts, personal communication, March 25, 2009). Residents of the West are unwilling to sacrifice the benefits they derive from their lands to benefit energy companies. The perception of inequity is particularly true regarding the energy industry among residents impacted most directly by intensified energy development.

There are pockets where conflict between energy developers and ranchers is more widespread (D. Simpson, personal communication, March 23, 2009). In New Mexico, animosity is relative to the personalities and belief systems of the persons interacting with each other, and, when agreements cannot be reached amicably, the conflict oftentimes gets "personal" and "emotional" as the issue "usually comes down to control of the land" (T. Herrell, personal communication, May 20, 2009). According to Tony Herrell, "85 percent of land management is emotion management." For example, Herrell notes that:

Usually, they [energy developers] get a landowners agreement. It's worked out. When it's not, a lot of times it'll be a surface owner saying, 'this company here can come in my land, or this company here can come in my land, but that company can't. Or sometimes it gets so personal that this person cannot come; your company is okay but these two or three, they can never step foot on my property again, and…what I've honestly come to believe is land brings out emotion in people, and control of the land is something that they've always had range-wars over, and we're still having modern-day range-wars, whether it be through split-estate or through the court system. The real issue is about control of the land and what happens on it. (T. Herrell, personal communication, May 20, 2009)

In controlling for the emotions of conflicting interests in how the land is going to be used and with whom the greatest control over those decisions is kept, according to Pat Shea, BLM administrators who are adept at balancing the interests and decision-making are essential. In Shea's opinion, if the political conflict between ranching and energy developers is to be avoided and the stability of the relationship within the BLM's land-use subgovernment maintained, BLM administrators must be "really, really good in terms of balancing out the interests and keeping everybody around the table" (P. Shea, personal communication, June 2, 2009).

In former Director Shea's opinion, balancing the interests in use and decision-making is lost when administrators favor one use over another. As Shea notes, "Literally in 2001, it [APDs] went from 800 for Sublette County [Wyoming] up to 2,000 and 4,000, and then in just one area the high number was 9,000 per year. So it just blew everything apart and when that happens the other responsibilities of both statutory and regulatory integrity go out the window. The feeling of the other con-stituencies be they ranchers or recreationists is, 'we don't count at all'" (P. Shea, personal communication, June 2, 2009). Shea's claim is supported, in part, by how some administrators interpret the dominance of the mineral estate in relation to the interests of surface owners. Generally speaking, all BLM administrators express concern with how surface owners are treated by industry representatives. Their con-cern, however, is tempered by the legal dominance of the mineral estate. As Don Simpson noted:

When the law tells us what's the dominant estate, then we write the regulations, we have to protect that, that dominant estate. I don't know if protect is the right word, but we have to honor it or recognize it. But what we also want to do is not have them run over the surface owners. What does the public think; what does the livestock, landowner think [about the fairness of the mineral estate's dominance over the surface estate]? All we want to know is did you [energy developer] make a deal with the landowner, and if they say 'yes,' we're good, either that or bond. (D. Simpson, personal communication, March 23, 2009)

Federal law and regulation of split-estate energy development favor energy develop-ment interests. If the balancing of ranching and energy interests is to be retained in land-use decision-making, the BLM must have the capacity to protect the interests of the surface owner as well as those of the energy developer. Currently, the scope of the BLM's protective oversight of split-estate surface properties is limited to pre-site inspection and post-development monitoring. Monitoring and enforcement beyond the federally mandated environmental protection and reclamation laws— most often addressed during the proposed energy development planning period— BLM oversight of split-estate energy activity is sporadic. Generally speaking, administrative oversight beyond what is required by federal law only occurs at the request of the surface owner.

Legal protection of the mineral estate's development as the preferred use of the land makes balanced administrative oversight of split-estate energy development difficult if not impossible. The role of a BLM administrator prior to energy activities taking place on split-estates is very limited. The assistance of a BLM administrator is most often requested only when energy activities have already begun and something goes wrong. This is problematic because problems that might have been addressed by the surface owner during the Surface Use Agreement negotiation, or in the pre-site inspection, do not benefit directly from BLM administrative expertise. In turn, when problems do occur, BLM administrators are held responsible for assisting in resolving the conflict that emerges between rancher and energy developer.

BLM Oversight and Federal Revenue

As domestic energy development expanded, administrative resources of the BLM strained to comply with their mandated oversight of energy activities. The preferred policy objective of the Bush Administration was to expand domestic energy development. BLM resources, however, did not increase as energy development expanded. As a result, while the number of acres under development increased to historical levels, BLM oversight of energy development activities declined. Having been directed by the President and his political appointees to expedite the APD approval process, BLM administrators found compliance with mandated oversight responsibilities difficult to achieve. In the opinion of BLM administrators, operating under the context of a political mandate to expand energy development, BLM lacked sufficient resources to expand administrative oversight of energy activities.

Disbursement and designated use of federal resources by federal agencies are matters for Congress. If a shift in policy direction is to take place, budgets must shift along with the objective being sought. According to Don Simpson:

> We shift all the time, but we don't shift the money. Congress gives us line items, so I would say in the last 10 years, or some period like that, [funding for] range [management] has gone down; recreation has gone down; oil and gas has gone up. Those were line items from Congress, they reprioritize our money, then our boss, the Secretary of the Interior, passes them down to the [state] director and says 'Here's the priorities,' and they [priorities] just kind of bounce around, so it depends on what's going on. (D. Simpson, personal communication, March 23, 2009)

Larry Claypool, Simpson's assistant director, clarifies that the shift in resources that occurred during the Bush Administration was "the shift, in the APDs, the big shift in the APDs, the major shift is probably our pilot offices[4] in that we hired additional

---

[4] Note: The pilot offices Mr. Claypool is referring to are offices within close proximity to fields where the greatest energy development activity is occurring. The BLM pilot offices are unique features of land management agencies in that they are devoted to no other administrative function other than that of oil and gas development.

people to take care of that extra workload in the permit area" (L. Claypool, personal communication, March 23, 2009). Shifting the administrative priorities of the BLM to expedite APDs had the effect of creating a backlog of regulatory compliance oversight in the field. In part, this is because agency budgets are created in years prior to any politically mandated shifts in policy priorities. There is a significant lag-time between BLM submitting a budget based on projected needs and requesting funding to meet immediate needs should a shift in administrative priorities occur.

Budget lag-time worked against BLM administrators' capacity to monitor energy activities. A June 2005 Government Accountability Office (GAO) noted that energy permitting activities tripled from 1999 to 2004 (Government Accountability Office Report (GAO-05-418), 2005). BLM permitting in 1999 accounted for 1803 APDs being issued. By 2004 the number of approved APDs numbered had risen to 6399 per year and was climbing. GAO noted that "BLM officials in five out of eight field offices that GAO visited explained that as a result of increases in drilling permit workloads, staff had to devote increased time to processing drilling permits, leaving less time for mitigation activities, such as environmental inspections and idle-well reviews" (Government Accountability Office Report (GAO-05-418), 2005, pg. 1). The report further noted that four of the eight BLM field offices "reported that the most significant impact of policies to expedite and manage oil and gas development was the increased emphasis that some of these policies placed on processing permits, which in turn resulted in shifting staff responsibilities away from mitigation activities" (Government Accountability Office Report (GAO-05-418), 2005, pg. 1). Thus, a change in the presidency and the support of a friendly Congress led to changes in BLM's management of domestic energy policy.

The BLM's response to executive branch directives emphasizing the expansion of domestic energy development altered the agency's administrative priorities. In response to Executive Orders #13211 and 13,212, APD backlogs and new APD application were being addressed by BLM administrators. In turn, a backlog in the monitoring and inspection of energy activities was created. While the executive branch had effectively shifted the energy policies of the BLM, Congress was slow to respond in allocating funds to balance BLM workloads. As Don Simpson notes, "Okay, so you've [Congress] got enough money here and you've told us this should account for some number of APDs. Well, guess what? It does, but as we add 5,000 more APDs per year to manage the compliance workload is going up by that amount. So we run back and say, 'Well, that's not enough money. You're funding the front part, but not the back part'" (D. Simpson, personal communication, March 23, 2009). The lack of funding was not, however, simply the result of the BLM responding to executive branch policy objectives. Congress too had a role to play in creating the administrative imbalance.

A Republican-dominated Congress reacted favorably to Bush energy policy objectives. According to Rebecca Watson:

> You have to remember how the federal government works. Congress, in the Constitution, is given authority over public lands and Congress also, of course, controls the budget. So, Congress and the White House were in concert in their belief that natural gas supply was

diminished and we needed more natural gas. Congress reacted by focusing on energy and the Energy Policy Act. The Bush Administration, from the very beginning was focused on energy and the need to supply domestic energy. And then the budget reflected that and the Bush budget drives policy and it's the budget that reflected the need for more money to develop natural gas and other energy and Congress passed those budgets. (R. Watson, personal communication, June 16, 2009)

Political control of both the BLM's administrative activities as well as their budget, in part, helps explain how the BLM shifted resources toward energy development activities. It is common that elected officials, and particularly appointed officials, understand controlling the budget means controlling the agency. As former BLM Director Pat Shea notes, "I came away from my experience in the department [Department of Interior] and in BLM absolutely convinced that the only way a political appointee can make a difference is by the control he or she took of the budget. You could make all sorts of administrative changes, and there would be temporal victories, but the real sustainable victories were the ones that you put into the budget" (P. Shea, personal communication, June 2, 2009). If a political appointee's control of an agency's budget sustains policy change, it can be surmised that in carrying out President Bush's executive orders, BLM's policy and budgetary priorities shifted away from ranching activities and toward energy development activities.

There is greater economic return to government on developing domestic energy resources than from ranching activities. While Rebecca Watson insisted that "There was no directive [to the BLM] to raise money," she also notes that as a result of market forces, government's economic gain from developing energy resources is "just the byproduct" (R. Watson, personal communication, June 16, 2009). According to Ms. Watson, the Bush Administration's message was not "go out and drill gas to raise money" because federal revenues from energy activities, when compared to other sources of revenue, are relatively minor. As Watson notes, "Yes, natural gas and oil and coal and other mineral resources bring in billions of dollars to the Federal Treasury, but that's a pimple compared to the money that's raised through taxes" (R. Watson, personal communication, June 16, 2009). While decision-makers were mindful of the impact of raising monies through energy development, DOI-appointed officials laid significant blame at the feet of the Office of Management and Budget (OMB) for the inequitable distribution of financial resources between ranching and energy. According to the senior DOI appointee:

You've got to remember, you had the Office of Management and budget that absolutely detested grazing on public lands. I mean, they do not like it. They never have. They don't think ranchers pay fair market value. So you've got an OMB that's going 'Fuck them. I'm not going to increase the budget for them. They're not paying fair market value. They want us to give them more money and to graze more and do more damage to the land.' So even though Interior would always ask for more money, the OMB would cut it back. (Unnamed DOI political appointee, personal communication, May 26, 2009)[5]

---

[5] Note: Fair-market value regarding monies paid by industry for energy leases averages $2.00 per acre.

Rebecca Watson also noted those problems with OMB, and the allocation of resources to the BLM for sustaining ranching activities is difficult to achieve. As Watson echoed, "I mean there's a whole other story about OMB and their role and what they do and who the people are at OMB and what kind of decisions they make on all manner of issues. That's a whole other debate" (R. Watson, personal communication, June 16, 2009). In some instances, according to the unnamed DOI appointee, OMB baulks at funding energy development as well. The appointee notes that "The Buffalo [Wyoming] field office was predicated on the fact that if you [BLM] give them [energy developers] more APD approvals, you'll [government] get more royalties, so OMB are you stupid? The state's [Wyoming] out there draining the hell out of you producing oil and gas from their state lands, and their draining the federal reserves, and your losing as much as 80 million dollars a year by not granting more APD reviewers to get these wells permitted so that you don't get drained by the state" (Unnamed DOI political appointee, personal communication, May 26, 2009). BLM administrators did not take a position on the OMB debate, but they too regarded energy market forces as a significant factor in creating resource disparity between ranching and energy. BLM administrators were, however, as mindful as their DOI counterparts about the desirability of raising federal revenues from domestic energy development.

BLM administrators are mindful of their role in raising federal revenue from energy activities. If energy companies can extract and develop energy resources when prices are high, elected officials seek to take advantage of the market price as a means of deriving revenue. As Lynn Rust notes, "Price is a big thing with it. So many people talked about well, the Bush energy policy. It's all about price. If companies can make money out there, they're going to go out there and drill and produce. If they can't, they're going to go elsewhere" (L. Rust, personal communication, May 19, 2009). Administrators like Rust are also mindful of the effect of elected officials seeking to pad the bottom line and make up for any budgetary shortfalls that might befall them in the future. As Rust comments, "There's a lot of revenue that they [federal agencies] know the federal government is dependent upon particularly in the current [2009] budget situation that's occurring, they're really looking carefully at it. They're [elected officials] looking for every dime they can find" (L. Rust, personal communication, May 19, 2009).

In states where energy resource development is greatest, such as Wyoming, the emphasis on the subsurface estate's capacity for raising revenue is particularly acute. For example, Larry Claypool notes that from the perspective of history, which estate derives more governmental revenue has changed. The surface estate is no longer viewed as the revenue producer it had once been. In Claypool's opinion:

> You [government] own the land. Poof [Stock-Raising Homestead Act of 1916]. The rancher owns the surface. The surface, it's there, but it just didn't carry the same weight in historical times as it does now. It just wasn't important. And it is interesting that back in the early 1900s the government saw at that time the start of the production of oil and gas. I was really surprised they [federal elected officials] thought this [federal government retaining ownership of the mineral estate] was a wise move; let's keep those minerals for the government, and kudos to them [federal elected officials] that they foresaw that and took the steps to put that [mineral estate] back in the government's hands. It was a good move. (L. Claypool, personal communication, March 23, 2009)

Currently, and for the foreseeable future, expanding domestic energy development will produce greater governmental revenues than will ranch activities. And, when the political objective is altered to take advantage of economic opportunity, the BLM shifts its policies as it responds to the political objective being sought. As Don Simpson notes:

> We have congressmen, we have senators, we have the president, and they all dictate through funding, through priorities, through executive orders, through laws, through regulations, how it is that we should behave. Well, and the forefathers reserved it [mineral estate] for all of us, and those that have passed laws since then said, 'Use it.' I mean, the laws mostly say 'use it.' They don't say 'hang on.' So, I think it's pretty clear that for a couple of hundred years that's kind of been the marching orders and we're [BLM] the intermediary, I guess, to stand back and step in when asked. (D. Simpson, personal communication, March 23, 2009)

In the modern energy economy, the political, legal, and administrative behavior of governmental entities will continue to favor the development of domestic energy resources over the economic interests of ranching. And as the BLM responds, the interests of energy within the subgovernment of the BLM will become further entrenched. That energy interests will be the focus of governmental entities at all levels of the federal government does not bode well for the future interests of the ranching community.

Conclusion: Disruption of a Subgovernment

Disruption of the BLM's land-use subgovernment was triggered by the Bush policy of expanding domestic energy development. In response, the BLM shifted their policies and resources in a manner that favored the interests of energy developers. The expansion of domestic energy development then spilled onto split-estate lands in a manner favoring energy interests. In turn, conflicts began to emerge as more split-estate lands were developed for their energy resources. Simply stated, as split-estate energy development multiplied, it triggered conflict and competition between ranching and energy.

As most government officials note, the dominance of the federally owned subsurface mineral estate in the context of federal land management is legally protected and problematic. Federal regulations guiding split-estate development reflect the legality of federal leasing and permitting of the subsurface. Due to the legal and regulatory protection of the mineral estate's dominance, animosity between ranchers and energy developers emerged. Thus, conflict over control of the land and its use established the conditions for a political conflict to emerge between ranchers and energy developers.

Multiple factors affect the development of energy resources. Thus, while disruption of the BLM's subgovernment is the result of executive branch actions, the conflict between ranching and energy is the result of where and how the energy is developed. In this case, the conflict between ranching and energy centers on split-estate energy development. The ability to access large, undeveloped, untraditional, energy resources such as coalbed methane (CBM) coincided with the unified political will of a Republican-controlled federal government and spilled onto split-estate lands across the Rocky Mountain West. In the rush to develop energy resources, the federal government enforced and defended the mineral estate's dominance over the privately owned surface estate. In turn, ranching interests wrestled with the new reality of energy development's ability to affect the politics of land-use decision-making.

While government officials recognize that "there are a lot of forces at work on grazing" (R. Watson, personal communication, June 16, 2009), the general consensus among this group of actors is that the legal dominance of the mineral estate and its exploitation for energy resources has had a detrimental effect on ranching's once-formidable influence on the BLM's subgovernment. The depth and breadth of power wielded by ranchers is illustrated by Rebecca Watson's comment, "They have a voice that is still listened to even though economically they don't play the same role. They play an important role in the West's culture. They have a strong voice, they wear the white hat. They mean something to a lot of people." There was also general consensus among government actors that ranchers could, should they choose to organize with other interests, gain the upper hand in reforming the dominance of the mineral estate in split-estate energy development (Interviews collectively).

Another dimension is that government officials also believe that energy developers have a difficult task made more difficult with "the complexities, the issues, all the things they have to consider" (L. Claypool, personal communication, March 23, 2009). Pat Shea notes that "there's no mutuality of economic interests" between ranching and energy, and in many ways this economic reality allows, even encourages, energy development to retain their upper hand within the BLM's subgovernment (P. Shea, personal communication, June 2, 2009). Finally, there is general consensus among the governmental actors that the political conflict that eventually emerged between ranching and energy could have been avoided if the trust between the two groups had not broken down (Interviews collectively). The conflict and competition between ranching and energy interests, addressed in the next two chapters, are explorations into how the interests strategized and deployed resources as they sought to control the land-use subgovernment of the BLM.

References

Claypool, L. (2009). Deputy State Director of Minerals and Lands Wyoming Bureau of Land Management. Interview Conducted: March 23, 2009; Cheyenne, WY.

General Accountability Office. (2005, June). Oil and gas development: Increased permitting activity has lessened BLM's ability to meet its environmental protection responsibilities (GAO-05-418).

Herrell, T. (2009). Deputy State Director of Minerals and Lands New Mexico Bureau of Land Management. Interview Conducted: May 20, 2009; Albuquerque, NM.

Roberts, E. (2009). Colorado State Representative (R-Dist. 59). Interview Conducted: March 25, 2009; Denver, CO.

Rust, L. (2009). Deputy State Director of Minerals and Lands Colorado Bureau of Land Management. Interview Conducted: May 19, 2009; Denver, CO.

Shea, P. (2009). Former Director of the Bureau of Land Management under President William J. Clinton. Interview Conducted: June 2, 2009; Salt Lake City, UT.

Simpson, D. (2009). State Director of Wyoming Bureau of Land Management. Interview Conducted: March 23, 2009; Cheyenne, WY.

Unnamed (2009). Former Senior Department of Interior Political Appointee under former President George W. Bush. Interview Conducted: May 26, 2009; Salt Lake City, UT.

Watson, R. (2009). Former Assistant Interior Secretary for Lands and Mineral Management under former President George W. Bush. Interview Conducted: June 16, 2009; Denver, CO.

Chapter 6
Energy Developers

Abstract Changes in domestic energy policy triggered heightened conflict and competition between the formerly allied, strong, and resource-rich members of the Bureau of Land Management's (BLM) public lands subgovernment: energy and ranching. From the perspective of energy industry representatives, the heightened conflict and competition with ranching resulted from variety of conditions. In the opinion of these industry representatives, expansion of domestic energy development was the result of political willpower, energy market forces, and technological advancements. There are, however, differing opinions regarding the extent to which split-estate energy development triggered conflict and competition between itself and the ranching industry. While there is general consensus among industry representatives that expanding split-estate energy development did impact the stability of the energy-ranching alliance, there is disagreement concerning the extent to which the alliance has been strained. There is also disagreement regarding how or why the conflict with ranchers became as heightened as it did as split-estate energy development expanded across the Rocky Mountain West. Representatives of the energy industry clearly believe that split-estate energy development triggered conflict with their ranching brethren.

Keywords Energy lobby · Bureau of Land Management · Department of Interior · Bureaucracy · Split-estate energy development · Surface Owner Protection Act

"…industry [energy] confuses the heck out of the BLM. I mean, it's government. It can't stay ahead of all that's going on." (Kathleen Sgamma, Director of Government Affairs, Independent Petroleum Association of Mountain States)

The Voice of Energy

Changes in domestic energy policy triggered heightened conflict and competition between the formerly allied, strong, and resource-rich members of the Bureau of Land Management's (BLM) public lands subgovernment: energy and ranching.

© Springer Nature Switzerland AG 2019
R. E. Forbis Jr., *Altered Policy Landscapes*,
https://doi.org/10.1007/978-3-030-04774-0_6

From the perspective of energy industry representatives across the states of New Mexico, Colorado, and Wyoming, the heightened conflict and competition with ranching resulted from variety of conditions.[1] In the opinion of these industry representatives, the expansion of domestic energy development was the result of political willpower, energy market forces, and technological advancements. There are, however, differing opinions regarding the extent to which split-estate energy development triggered conflict and competition between itself and the ranching industry. For example, while there is general consensus among industry representatives that expanding split-estate energy development did impact the stability of the energy-ranching alliance, there is disagreement concerning the extent to which the alliance has been strained. There is also disagreement regarding how or why the conflict with ranchers became as heightened as it did as split-estate energy development expanded across the Rocky Mountain West.

Representatives of the energy industry clearly believe that split-estate energy development triggered conflict with their ranching brethren. Their opinions, however, are mixed concerning how the conflict happened. They have equally varied opinions regarding the degree to which the conflict has impacted the energy industry's control of the BLM's land-use subgovernment. Some industry representatives are of the opinion that problems associated with split-estate energy development are complex, dynamic, and troubling. They also share the opinion that the conflicts created by those problems are infrequent, emotional, and exploited by other interests. Finally, there is no agreement among industry representatives over the question of whether, as a result of their conflict with ranchers, their industry achieved dominance of the BLM's land-use subgovernment away from ranching.

The voices represented here are those of energy industry elites who offer their perspective on the conditions that led to conflict with the ranching industry. They are defined as elites because they are a select sample of actors who interact regularly with other interest groups and government entities that compose the networks of the BLM's land-use subgovernment. These elites were active representatives of their industry's interests at the height of the energy industry's conflict and competition with ranching.

---

[1] Energy Industry participants include:

(1) Kathleen Sgamma, Director of Government Affairs for the Independent Petroleum Association of Mountain States (IPAMS)

(2) Bob Gallagher, former President of the New Mexico Oil and Gas Association (NMOGA)

Note: Bob Gallagher was dismissed from his position following his 2009 interview (confirmed by NMOGA on January 28, 2010). His dismissal was not the result of comments made during the course of the interview as neither the recorded interview nor its transcript has ever been made public before now.

(3) Stan Dempsey, President of the Colorado Petroleum Association (CPA)

(4) Bruce Hinchey, President of the Petroleum Association of Wyoming (PAW) and former Speaker of the House, State of Wyoming Legislature

Shifting the BLM's Energy Policies and Resources

Conflict and competition among interest groups interacting within a subgovernment is not unusual. What is unusual, however, is when the interests of two elements of a historical alliance within the subgovernment collide. In the case of the BLM's land-use subgovernment, the historically allied interests of the energy and ranching industries collided as the BLM shifted its policies and resources away from ranching and toward energy. The shift in the BLM's energy policy was, in part, politically motivated. As was previously discussed, energy markets and technological advancements in the extraction of non-traditional energy resources also played a significant role in shifting the BLM's energy policy. According to Bob Gallagher, former president of the New Mexico Oil and Gas Association, "Just previous to the 2001 [Bush] executive orders, the eight years of the Clinton Administration were tough, tough years for the industry. [After the 2000 election of Bush] there was a more determined effort to see the domestic oil and gas industry come out of the ashes, and come out strong, and [energy development] just needed the political will, and that was provided by the administration" (B. Gallagher, personal communication, May 21, 2009). Gallagher further notes that prior to the 2000 election, energy markets as well as technological advancements aided in expanding domestic energy development, and he refers to the combination of those conditions as "the perfect storm" (B. Gallagher, personal communication, May 21, 2009).

Other industry representatives do not necessarily agree with the assessment of the scenario as a "perfect storm." Other industry representatives discount the effect of President Bush's political willpower on the expansion of domestic energy development. According to Kathleen Sgamma, Director of Government Affairs for the Independent Petroleum Association of Mountain States:

> There was certainly was an effort by the administration to encourage production of domestic resources. [But], in no way do I think that was a catalyst for what occurred with the growth of industry. You saw in the late 1990s the ability with technology to start going after unconventional reserves that we weren't able to go after before. Coupled with that [were the] rising commodity prices that occurred, we had a converging of forces. That technology came online at the same time that commodity prices were taking off. So, the catalyst was not the Bush Administration issuing that executive order, because, you know, you can order an executive order all you want. What company is going to spend millions of dollars to lease lands and go after something if they don't have the technology or they don't have the price to make it economic? So, I mean, certainly the Bush Administration encouraging domestic production was helpful to a certain point, but it was really those forces. They're [energy companies] not going to put money in if they're not getting money out. I mean, nobody wants to drill holes for the point of drilling holes. (K. Sgamma, personal communication, March 24, 2009)

Stan Dempsey, president of the Colorado Petroleum Association, suggests that President Bush's political willpower was more akin to the political messaging one receives with every electoral change in administration. The Bush message was clear. According to Dempsey, there was certainty in that, "The Bush Administration had a different view of extraction and resources" (S. Dempsey, personal communication, March 24, 2009). In Dempsey's opinion, like that of Ms. Sgamma's, "two factors

[were] very important" and more directly related to the expansion of domestic energy development: energy markets and technological advancements (S. Dempsey, personal communication, March 24, 2009). And finally, according to Bruce Hinchey, president of the Petroleum Association of Wyoming, President Bush's political will-power had nothing to do with the expansion of domestic energy development. In Mr. Hinchey's opinion, the Bush Administration was more onerous in "attacking" and "regulating" the energy industry than had been the Clinton Administration. According to Hinchey, "I thought well here is the Bush Administration supposedly opening up the gates and yet, we're [industry] seeing all these regulations just coming out of ears. Stips [stipulations] dealing with more reclamation stips, bird stips, you name it. Winter drilling stips [were] just coming out our ears during the Bush Administration" (B. Hinchey, personal communication, June 12, 2009).

According to most energy industry representatives, President Bush's executive orders did not have the effect of shifting the BLM's energy policies or procedures. In their collective opinion, if any shift occurred, it was in the allocation of BLM resources to meet the needs of the energy industry as development of non-traditional energy resources expanded. According to Mr. Hinchey, the BLM "added people because they had to get out more Approved Permits to Drill (APD)" (B. Hinchey, personal communication, June 12, 2009). He adds that "They [BLM] couldn't do it. So we [industry] helped fund that" (B. Hinchey, personal communication, June 12, 2009). As discussed previously, the intent of President Bush's executive orders was to expedite the APD process. But, according to Mr. Hinchey, while the number of APDs being issued rose steadily to an average of around "2500" per year, the issuance of APDs "never hit their 90-day" regulated deadlines. In some instances, "they [APDs] were 180 days even after the executive orders came out" (B. Hinchey, personal communication, June 12, 2009). Consequently, according to Hinchey, "they [Presidents] can issue all the executive orders they want. It doesn't mean anything" (B. Hinchey, personal communication, June 12, 2009). Thus, the impact of a President Bush's exercise of unilateral executive powers on the BLM's energy policies is disputed among representatives of the energy industry.

Industry representatives believe that energy markets and technological advancements were more directly related to growth in the energy industry. In turn, industry representatives are of the opinion that the BLM was hard-pressed to meet industry's demand for APDs as energy developers sought to take advantage of favorable market and technological conditions. The phenomenon of domestic energy development expanding across the Rocky Mountain West is then analogous to a row of falling dominoes. As the price of energy rises, the requested number of APDs increases, and as the number of APDs issued increases, energy development activity expands across the West. In turn, BLM administrative resources strain to meet the rise in energy development activity. Therefore, any redistribution of the BLM's administrative resources shifts when the procedural requirements of energy development activity increase. As Bob Gallagher notes, "The proof is in the pudding. In January of 2000, the first couple of years were dominated by just battling the BLM. There were 145 day waiting periods for APDs to be approved and you have to do this, and you have to that. Now all that has changed, in two of the largest out of the top four top oil and gas offices in the United States of the BLM, we are still paying for the archeological surveys and reports from a third-party and handing them in because the BLM can't do it" (B. Gallagher, personal communication, May 21, 2009). Consequently, government resources, even when distribution of administrative resources shifted, remained inadequate for the BLM to fully address their mandated procedural and regulatory oversight of industry activities as domestic energy development expanded across the Western United States.

Expanded Development of CBM Energy Resources

Non-traditional energy resources became economically feasible for industry to develop (K. Sgamma, personal communication, March 24, 2009). From the perspective of industry, the growth of coalbed methane (CBM) development across Western states was not a response to political willpower; growth was a response to market demand and the technological capacity to inexpensively develop CBM energy resources. Consequently, growth in CBM energy development was rapid. As Stan Dempsey, president of the Colorado Petroleum Association, notes, growth in CBM activity is the result of "technology and the ability the companies have had to develop in tight sand formations and the ability to drill up to 30–33 wells off a single pad, to develop this kind of resource" (S. Dempsey, personal communication, March 24, 2009). Prior to the technological capacity of the energy industry to extract CBM as a resource, CBM was considered by industry as too expensive to develop. Therefore, when the technological challenges of developing CBM had been overcome, CBM became an inexpensive and highly sought after new energy resource.

In the late 1990s, the technological capacity to develop CBM became commonplace among energy developers, and CBM activity spread rapidly. This occurred because, over time, CBM emerged as a new and plentiful energy resource. The energy industry's newly increased technological capacity allowed it to develop and market CBM. As a result, CBM activity spread rapidly across states of the Rocky Mountain West. This is because some of the largest and most easily accessible fields of underground coal seam formations—where CBM is found—are located within the states of New Mexico, Colorado, and Wyoming. As CBM development grew, pockets of intense CBM energy development appeared across the states of New Mexico, Colorado, and Wyoming. As Bruce Hinchey, president of the Petroleum Association of Wyoming, notes, "We have pockets of development. If you look at the Pinedale/Jonah [Wyoming] Area, it's around 30,000 acres, and then you're gonna' drive 60, 70 miles to Rock Springs [Wyoming] and not see a well. So, there's pockets of development" (B. Hinchey, personal communication, June 12, 2009).

As more CBM became available for development, split-estate energy development intensified across the West, increasing interaction between energy developers, the surface-owning ranchers, and homeowners of the West. As discussed in the previous chapter, the interaction between these two groups was not unusual as most Western landowners, particularly old-school ranchers, are familiar with energy development. This time, however, the level of energy development activity surpassed anything even those most familiar with previous energy booms had ever experienced. The effect was that interaction between energy developers and surface owners increased as levels of split-estate energy development increased. Often, the interaction was with surface owners who were unfamiliar with energy development. In either case, conflict between energy developers and surface owners intensified. As Bob Gallagher notes, "It [conflict] then obviously ballooned into something that I don't think the BLM was ready for, and something that I don't think any of us thought truly would occur that quickly where all of the sudden you were having conflicts with other users of public and private lands, and it happened overnight" (B. Gallagher, personal communication, May 21, 2009).

Within pockets of intense split-estate CBM energy development, conflict between energy developers and surface owners was especially intense. As Mr. Gallagher describes the spread of conflict, it was as if a fuse had been lit, and as the fuse burned, it ignited animosity and distrust between energy developers and ranchers across the West. Conflict over energy development booms is not unusual, but in the case of CBM development, the growth of the conflict was different. As Gallagher suggests, "most of the time [controversy] will maybe start in New Mexico and the next year or two will trickle to Wyoming and trickle to Colorado, or vice versa," but, according to Gallagher, "in this case it didn't" (B. Gallagher, personal communication, May 21, 2009). Gallagher adds that, "All of a sudden, the entire West seemed to be engulfed, if you will, in that problem and how slow it took people to respond, how arrogant both sides were. Certainly it led to a prolonged period before anybody was really ready to collectively to sit down and figure out what we [energy developers and ranchers] could do" (B. Gallagher, personal communication, May 21, 2009).

As split-estate CBM energy development expanded, the growth of conflict was difficult for energy companies to manage. In the relationship between ranchers and industry, Dempsey notes that the intention of industry was to manage the conflict as split-estate energy development grew. In Dempsey's opinion, the growth of the conflict with ranchers can be attributed in part to:

> Some companies grew very quickly. I'm not going to say they grew too quickly; they grew very quickly, and there are companies out there who have strong public affairs who acknowledge that they grew very quickly and they had a difficult time, and I use the word manage. I think there have been companies in Colorado that as they ramp up their activity, there isn't a concerted effort by the company to best manage the whole scope of their operations. There have been some well-documented cases of perhaps companies running roughshod, but I think that's when companies are growing quickly and they don't have complete control over all aspects. The biggest problem sometimes, I think, is the relationship in management of their subcontractors. That's probably the best way to look at it. Are the companies really exerting and supervising, and it's not just treatment of the surface owner, it's the neighbor, it's, gosh, these trucks are driving, they ran me off the road, [they] killed my sheep. (S. Dempsey, personal communication, March 24, 2009)

As split-estate energy expanded across the West, energy companies not only struggled to contain conflict with ranchers, but they also struggled to manage their own growth. In turn, conflicts between energy developers and split-estate property owners multiplied. Conflict is, according to Dempsey and other industry representatives, a result of industry's inability to manage its growth. Or, as Dempsey notes, conflict is a result of some companies' inability to exert control over the behavior of their representative land-men or subcontractors.

The Federal Mineral Estate's Legal Dominance

The legal and regulatory dominance of the federal mineral estate influenced the behavior of energy developers toward split-estate property owners. Within pockets of split-estate energy development where development is most intense, conflicts between energy developers and landowners have been "hotter" (S. Dempsey, personal communication, March 24, 2009). Furthermore, underlying the heightened conflict is the dynamic interaction between energy companies and ranchers. The interaction is shaped equally by industry's reliance on the dominance of the federal mineral estate and ranchers' dislike of the mineral estate's dominance. Simply stated, the legal dominance of the federal mineral estate affects the negotiation process of Surface Owner Agreements. As Stan Dempsey notes, "I'm sure there are a lot of surface owners that aren't jumping up and down, thrilled to death, especially if they don't own the minerals, to have oil and gas activities on their land" (S. Dempsey, personal communication, March 24, 2009).

Successfully negotiating a Surface Owner Agreement with split-estate property owners is dependent on a number of factors. Among industry representatives there is general consensus that the more knowledgeable landowners are regarding the severed nature of their property, the greater likelihood they understand that energy development is a possibility. Similarly, the more experienced landowners are with energy development, the greater the likelihood of their having better information and understanding the process of negotiating a Surface Owner Agreement. According to energy representatives, landowners who lack similar knowledge of the full nature of the mineral estate's dominance, or have never experienced energy development on their surface estate, are often caught off-guard. According to Kathleen Sgamma:

> They're [energy developers] used to dealing with the issue [Surface Owner Agreement] on a regular basis and your mom and pop rancher all of a sudden gets a knock on the door in a new situation, right off the bat, it's something that they [rancher] don't know much about, perhaps, or they don't understand what's all going to happen, so, the situation would start off, forget all the lawyers, there's a learning curve that anyone would have to kind of get over. But, in general companies will sit down with the owners, their land-men will go out, or whoever they [company] have designated, sit down and arrange something and come up with an agreement. They'll explain the situation and come up with an agreement, come up with monthly or yearly payment, whatever they kind of work out, you know, you really don't hear much about it. (K. Sgamma, personal communication, March 24, 2009)

Knowledge of one's property rights and the land's potential for energy development activity is the responsibility of the landowner. Bruce Hinchey argues that Surface Owner Agreements regarding the potential disturbance of energy activities on the surface estate must be worked out and ranchers must be compensated for the disturbance (B. Hinchey, personal communication, June 12, 2009). But Hinchey also argues that energy companies should not be held responsible for a split-estate property owners' lack of understanding of an energy company's ability to access and develop the mineral estate (B. Hinchey, personal communication, June 12, 2009). In Mr. Hinchey's opinion, "I gotta' say, you gotta' be really stupid if you don't know that you don't own the minerals because anybody that owns land knows up front whether or not they own the minerals. Unless you're out in Neverland or came from Planet X and moved to Wyoming and don't understand that, then maybe you didn't know that. But, I think anybody that's a savvy rancher knows whether or not they own the minerals" (B. Hinchey, personal communication, June 12, 2009). The assertion is that there are two groups of split-estate property owners with whom energy representatives must negotiate: those who have owned their land for extended periods of time and are intimately familiar with the legal dominance of the federal mineral estate and those who have neither knowledge of the mineral estate's dominance nor of their property's potential for energy development activity.

Informed or not, split-estate landowners opposed to energy activity occurring on their property must acquiesce to the entry of the energy developer. Bruce Hinchey argues that even among the better informed and experienced split-estate property owners, there are those who flatly oppose the entry of energy companies onto their lands (B. Hinchey, personal communication, June 12, 2009). Referring to surface owners attempts at prohibiting access to the mineral estate, Hinchey states, "That's not gonna' happen" (B. Hinchey, personal communication, June 12, 2009). According to Hinchey, opposition to access and development results, "Because it's private property, it's their surface, they'd never had oil and gas development on it, and they didn't want anybody on their land. And yet you've got the mineral estate that is dominant and it has to be dominant. Otherwise you couldn't get in there and develop your minerals and it would be a taking [compensable taking of property by government under the 5th Amendment to the Constitution]" (B. Hinchey, personal communication, June 12, 2009). Kathleen Sgamma adds that:

> I think the split-estate issue; it's obviously a tricky one because the mineral rights take primacy. So, nobody wants, you know, there's a concept of property ownership that 'this is my land'; I can do whatever I want, and people don't often understand the whole concept of the severed rights. It's a situation fraught with, you know, the potential for conflict. Nobody wants to be told, 'well, you only own the property to a certain point.' (K. Sgamma, personal communication, March 24, 2009)

According to Sgamma, the dominance of the federal mineral estate and the ability of the industry to access and develop energy serve the greater good. In Sgamma's opinion, "But, if you think about kind of a greater good, these are minerals owned by all Americans, so to me the law is fairly sensible in that those mineral rights take primacy over the surface" (K. Sgamma, personal communication, March 24, 2009).

Landowners' displeasure with the federal mineral estate's dominance is tempered by the energy industry's insistence that companies amicably negotiate with split-estate property owners. According to the majority of industry representatives, conflict with split-estate property owners is rare. There is a general consensus that conflicts did arise, but that the causes of those conflicts was the result of numerous variables. For example, in Colorado, according to Stan Dempsey, the urban interface of housing development with energy activity was a primary cause for conflict (S. Dempsey, personal communication, March 24, 2009). This opinion is shared among industry representatives. Colorado, in comparison to New Mexico and Wyoming, is where the urban interface of small ranchettes and energy development was greatest (Interviews collectively). Dempsey adds, "It was more people buying their 35 acres and never knowing that they had minerals underneath their estate. Yeah, we certainly have people coming in and complaining about, you know, their farms being impacted but not as much. This last go-round, I mean, this decade has all been about the ranchettes" (S. Dempsey, personal communication, March 24, 2009).

In New Mexico and Wyoming, conflict occurred primarily between energy developers and ranchers. In New Mexico, according to Bob Gallagher, the primary cause for conflict was ranchers' emotional response to the rapid growth of energy development (B. Gallagher, personal communication, May 21, 2009). In this case, conflict over split-estate energy development was the result of the *rural* interface of farming and livestock with energy activity. In Wyoming, according to Bruce Hinchey, the primary cause of conflict was a combination of "Old, long time big ranchers, some other people that had moved into the state that were pretty wealthy, [as well as] environmental activists" (B. Hinchey, personal communication, June 12, 2009). Industry representatives as a whole conclude that among the primary causal variables creating conflict between their industry and ranchers was the behavior of individual participants: energy developers and ranchers alike.

Energy Developers and Ranchers: Stories from the Field

Conflict heightened as energy developers and ranchers interacted, due in part to the interaction of land-men unfamiliar with the cultural traditions of Western ranchers. On the flip side of that coin, ranchers unfamiliar with the fast-paced culture of energy development clashed with energy company land-men. As Bob Gallagher notes, politics, the energy market, and technological advancements aside, the clash of cultures heightened the conflict between energy and ranching. Gallagher comments that:

> [It was] the attitudes on both sides. I mean, the attitude of oil and gas firms saying, instead of saying 'good morning Mr. Jones, how are you?' It was 'you need to know the mineral estate is dominant over the surface estate and, you know, blah, blah, blah, and then on the other side having the show up at the gate with a gun and just saying, 'I don't need to talk to you.' So I think it was actually what I like to term as bad actors on both sides, and it's unfortunate because although that may have only been at that time 5 or 10 percent, those [conflicts] are the ones that got the [press] coverage. (B. Gallagher, personal communication, May 21, 2009)

Mr. Gallagher's perspective is shared among his energy industry colleagues. For example, Stan Dempsey notes that negotiations between land-men and ranchers rarely failed to amicably reach Surface Owner Agreements (Interviews collectively). Dempsey comments that "very rarely do companies throw their hands up and say 'I can't reach a Surface Use Agreement. I'm going to post a bond'" (S. Dempsey, personal communication, March 24, 2009). Dempsey further comments that "this is a long-term relationship that's going to exist between the operator and the surface owner, and you'll find in many cases the company spending more, a lot more than they have to, to move a road or to provide some kind of support for some activity completely unrelated to what is in the Surface Use Agreement to accommodate the needs of the surface owner" (S. Dempsey, personal communication, March 24, 2009). As Dempsey and others note, the reputation of the industry and its relationship with ranchers is at stake. Therefore, any accommodation required of an energy developer in order to satisfactorily reach an agreement with ranchers is of paramount importance. Subsequently, representatives of the energy industry attempt to manage the behavior of their land-men toward split-estate property owners.[2]

The energy industry, however, holds an advantaged position in negotiating Surface Owner Agreements. This is because split-estate property owners often lack adequate information and knowledge of an energy company's ability to lease and access their land. As a consequence of this lack of knowledge, landowners are often disadvantaged in the negotiation process because they cannot adequately address accommodation or compensation for energy-related activities. And, as most industry representatives note, landowners' collective lack of knowledge allows bad actors within the industry to behave badly toward landowners. Most split-estate property owners "strike the best deal possible and still aren't happy about it, but sometimes that's often the result of good negotiation" (S. Dempsey, personal communication, March 24, 2009). As industry representatives are quick to point out, "there are very sophisticated surface owners," but they also add that no matter the level of sophistication "surface owners should avail themselves of the best legal counsel that they can obtain" (S. Dempsey, personal communication, March 24, 2009). This should be done, according to industry representatives, in order to avoid the possibility of being treated poorly by a bad actor energy company (Interviews collectively). As Bob Gallagher notes, energy companies have "gone from a handshake of 'yeah, we'll grade your road to the barn in exchange for this, to legal documentation as it's being dictated by, the land man hands it to you, and if you've got a question he says, 'well, you've got to call the attorney in Houston or you've got to call the attorney in Denver'" (B. Gallagher, personal communication, May 21, 2009). Gallagher also notes that as a result, "[industry] did a good job of alienating the people we should have actually been talking to" (B. Gallagher, personal communication, May 21, 2009). Gallagher further comments that, "there [are] companies that get it, who absolutely understand and continue to have great relationships with landowners" (B. Gallagher, personal communication, May 21, 2009). Thus, it is the perspective of energy representatives that most energy development companies amicably negotiate Surface Owner Agreements with landowners. From the perspective of the energy industry, it is rare when a company does not negotiate fairly with a landowner.

---

[2] Note: The BLM does not report the number of times bonds are posted when Surface Owner Agreements are not reached.

The relationship between energy developers and ranchers is driven by two dynamics: energy companies who treat surface owners well and those companies who don't treat them well. Energy companies that earn the reputation of behaving badly are, according to most industry representatives of small, out-of-state, companies, whose land-men and subcontractors have no understanding of the ranching culture of the West. They are, as most industry representatives note, those companies or persons least familiar with the energy industry's long tradition of working with ranchers to reach a "handshake deal across the kitchen table" (Interviews collectively). As Bruce Hinchey argues, "Yeah, a lot of the cases it's because they, depending on who you're dealing with you can get these Texas boys that come up here and just try to run roughshod over ya'. And then pretty quick they get into it with a few ranchers and they find out that's not how it works" (B. Hinchey, personal communication, June 12, 2009). The implication among industry representatives was that the quick lesson learned by those who clashed with ranchers was learned at the end of a gun or, if that failed, in a court of law. It was not often that a company's land-man or its subcontractors would be met with the threat of armed confrontation. But, as Bob Gallagher and Bruce Hinchey both confirm, ranchers would, from time to time, actually pull the trigger a time or two (B. Gallagher, personal communication, May 21, 2009; B. Hinchey, personal communication, June 12, 2009). More often than not, incidents of armed confrontation were avoided.

Industry representatives acknowledge that bad behavior among individuals or individual operators within the energy industry does exist. The shared opinion among energy representatives is that the truly horrible incidents are "very rare." And representatives acknowledge that when the bad incidents occur, those incidents "give the industry a 'black-eye' in terms of their relationship with landowners and the public" (Interviews collectively).

The same bad actor behavior is found among "recalcitrant landowners" (S. Dempsey, personal communication, March 24, 2009). According to industry representatives, these bad actors are individuals who seek to punish energy companies by denying access as a means of extorting money or beneficial treatment from the company. Bruce Hinchey argues that the bad actor scenario is a two-way street in that "you also have ranchers that absolutely don't want you on their land. And it doesn't matter what you pay 'em, it's not gonna' be enough. So it goes both ways" (B. Hinchey, personal communication, June 12, 2009). Among this group Hinchey suggests there are factions, stating there is "a faction of people that wanted huge amounts of money for the Surface Use Agreements and for whatever action was gonna' be taken on their ranch" and "there was others that the money wasn't, had nothing to do with it, it was just a matter of we don't want you on our land, period" (B. Hinchey, personal communication, June 12, 2009). Another faction among ranchers is a group who, according to Bob Gallagher, "[just] want enough information to be able to assess the impact, short-term and long-term, on their property" (B. Gallagher, personal communication, May 21, 2009). Gallagher adds that energy companies, and the industry as a whole, could benefit from once again developing a working relationship with ranchers by sharing their expertise and information. Gallagher asks, "You [land-man] have all this information, okay? So if you have this information, why aren't you sharing it with the people who are actually going to be impacted by this" (B. Gallagher, personal communication, May 21, 2009)? In Gallagher's opinion:

> If this would have happened early on, if we [industry] truly would've understood that 95 percent of those landowners that we're dealing with truly just want information to assess the impact, I think it would've gone a lot smoother. There's going to be some ranchers out there and they're going to make a living off of us. It happens all the time. I mean, if you killed, unfortunately, if 10 cows died from one ranch, and they have a thousand head of cattle, I can promise you all 10 of those were the blue-ribbon winners at the last county fair. But, you're going to have that on either side. But I truly think that is they [ranchers] were given, if they were communicated with from the start, and given the information they needed to do a true assessment of the impact on their land, I don't think we would have had any problems. (B. Gallagher, personal communication, May 21, 2009)

The incidents in which energy developers clashed with ranchers were the exception to the rule as energy developers and ranchers interacted. As has been noted, incidents where conflict occurred were the result of a variety of underlying factors. Most often those conflicts were driven by "bad actors" (Interviews collectively). Bad actors, according to energy representatives, were distributed equally among energy companies and split-estate property owners. Industry representatives share the view that bad actors exist among large, multinational energy companies, as well as small, independent, energy companies. And, as discussed previously, industry representatives also share the view that among split-estate property owners, there exist "bad actors" as well. Additionally, there is also a shared belief among energy representatives that managing conflict is an essential element to sustaining industry's relationship with ranching.

These assessments, descriptive as they are, do not fully explain how conflict and competition between energy and ranching interests heightened. If incidents were as rare as has been described, why did ranchers, often collectively, seek the assistance of their respective state legislatures in defending their interests from the harm of energy development? It is the shared opinion among industry representatives within the states of the Rocky Mountain West that the rare egregious incidents were exploited by outside interests seeking to divide the energy-ranching alliance: environmentalists.

The Intervention of Environmentalists

The conflict and competition between energy development and ranching interests would not have heightened had it not been for the interference of environmental organizations. According to energy representatives, the conflict between energy and ranching interests heightened because the conflict itself "started to be fueled by East Coast liberal organizations that have pumped millions on millions on millions of dollars into the West" (B. Gallagher, personal communication, May 21, 2009). According to a majority of industry representatives, as problems associated with split-estate energy development began to gain publicity, environmental organizations exploited those problems to create and heighten conflict and competition between energy and ranching interests. According to Bruce Hinchey, as ranchers began seeking the assistance of state legislators and demanding the enactment of protective legislation—referred to as Surface Owner Accommodation Acts[3]—"You had environmental activists that was really pushing that you never had before" (B. Hinchey, personal communication, June 12, 2009). It is the shared opinion among all, but one energy industry representative the intervention of environmental activists was the primary cause for heightening industry's conflict and competition with ranching over Surface Owner Protection Acts. Hinchey adds that:

> Absolutely, there's no doubt about it. They [environmental activists] were orchestrating and organizing and they still are and, it's way different than it was back in the '80s. You didn't see that. You didn't have those groups out there like that. And, they were really orchestrating and pushing this to, as much as they can and it's not just because they were activated out of Wyoming because they're activated out of a national network. That's what caused a lot of it [conflict and competition]. It's associated with the environmental movement. And, those organizations have thousands of oil and gas members and most of our oil and gas members don't even know that they're [environmental organizations] doing that. It's all a coordinated effort because if you look back into where their [environmental organizations] meetings were and who they were talking to, it's been the environmental extremists and they were pushing it [conflict and competition]. (B. Hinchey, personal communication, June 12, 2009)

And, as Bob Gallagher adds, "We don't need an East Coast liberal organization telling us [energy and ranching] how we're going to live or do business in New Mexico" (B. Gallagher, personal communication, May 21, 2009). Kathleen Sgamma echoes these sentiments when she argues that "I think they've [environmental organizations] exploited that [industry-ranching conflict] to hammer industry. They've certainly found a way to get in with landowners, disgruntled landowners and raise that [legal dominance of the mineral estate] as an issue. [To] just raise the issue [split-estate energy development] as if this was a huge problem, as if the majority of people were being walked all over by big oil and gas. So I think they've been pretty successful in exploiting that" (K. Sgamma, personal communication, March 24, 2009).

The belief that East Coast liberal environmental organizations heightened the energy-ranching conflict is not supported by the pattern of conflict in Colorado. The difference in Colorado, according to Stan Dempsey, is that while environmental organizations did participate during the legislative debates over Surface Owner Protection legislation, the primary cause for heightening the conflict was the intervention of homebuilder associations. Dempsey notes that while there was a short-lived alignment of commercial and housing developers with environmental

---

[3] Note: Surface Owner Protection Acts, or Surface Owner Accommodation Acts, were addressed or introduced in the states of New Mexico, Colorado, Wyoming, Montana, and Utah (2000–2009). In the State of Montana, existing legislation was reformed. In the State of Utah (2009), legislation was introduced but failed in committee. The States of New Mexico, Colorado, and Wyoming passed differing versions of the Acts.

organizations during the course of the debate, the intervention of environmental organizations was tempered by the fact that "homebuilders are inherently more conservative and had no interest in getting involved in what was viewed as a complete attack on the industry in Colorado" and "because homebuilders and environmental communities don't like each other at all" (S. Dempsey, personal communication, March 24, 2009).

Environmental organizations exploited the conflict with the intent of dividing the energy-ranching alliance; according to industry representatives, heightening the conflict between the energy and ranching industries serves the purpose of environmental organizations to "divide and conquer" their respective industries (K. Sgamma, personal communication, March 24, 2009; B. Gallagher, personal communication, May 21, 2009; B. Hinchey, personal communication, June 12, 2009). This argument suggests that ranchers are being duped into aligning themselves with environmental interest groups bent on ranching's destruction. As Kathleen Sgamma notes:

> I think a lot of ranchers; certainly some ranchers have joined forces in that respect. A lot of ranchers are very wary of environmental groups because they see them as, 'all right, we'll get rid of the energy companies and then we'll go after the ranchers.' You know, quite frankly, there's plenty of history of environmental groups trying to go after grazing allotments and drive them [ranchers] off. So, I think there's a built in mistrust, so sometimes these alliances get a little overblown. You see that with hunting organizations as well. It's hard to argue that something like a TRCP [Theodore Roosevelt Conservation Partnership] or a Trout Unlimited is truly just a hunting organization; they're more of an environmental organization. You know, their agenda is to really stop oil and gas. They want us out of here. They don't want to deal with energy development. So anything that they can do to stop energy development and anybody they can team up with to make that happen is great with them. I think some ranchers are, again, wary of those types of organizations. There are groups that clearly have an agenda of let's get rid of coal mining and then let's get rid of oil and gas and then let's get rid of the ranchers. I think sometimes the ranchers see through that. (K. Sgamma, personal communication, March 24, 2009)

The alignment of ranchers with environmental organizations is worrisome to the energy industry. Ranching's alignment with environmental interest groups is, among energy representatives, faulted with driving mistrust between energy developers and ranchers. The ranching-environmental alignment is also faulted with heightening the conflict between energy and ranching as the two interest began to debate the merits of enacting Surface Owner Protection Acts (K. Sgamma, personal communication, March 24, 2009; B. Gallagher, personal communication, May 21, 2009; B. Hinchey, personal communication, June 12, 2009).

Despite industry's failure to manage problems associated with the growth of split-estate energy's development, the attitude of the industry's behavior and interaction with landowners, the individual carelessness of bad actors, or the exploitation of environmental organizations, it is clear there exists a heightened level of animosity and mistrust between energy developers and ranchers. And, as the debate over the enactment of Surface Owner Protection Acts took place, mistrust drove animosity. It is mistrust and animosity that drove split-estate landowners, ranchers, and homeowners alike to their state legislatures. As will be discussed, the hierarchical relationship between energy and ranching within the interest network of the BLM's land-use subgovernment has been altered by those legislative battles over Surface

Owner Accommodation Acts. Simply stated, as legislative debates ensued across the states of the Rocky Mountain West, it was clear that the formerly allied interests of energy and ranching had grown wary of each other.

Surface Owner Protection Acts: Energy's Perspective

Frustrated and angry split-estate property owners sought the enactment of Surface Owner Protection Acts. In doing so, property owners sought the attention of their state-elected officials. As legislation was introduced, split-estate property owners sought the assistance of organizations in which they were members. And because the most powerful individuals among them were ranchers or wealthy landowners who ran small agricultural operations, landowners sought the lobbying assistance of state Stockgrower associations. In turn, state petroleum associations similarly began to lobby state-elected officials in opposition to the proposed Surface Owner Accommodation Acts. The elite representatives of the energy and ranching lobbies began to engage in a battle for control of land-use.

Conflict and competition heightened as energy and ranching organizations clashed over state legislation. The interest groups clashed because, as discussed previously, the BLM could not adequately respond to problems associated with split-estate energy development. And because the BLM could not always resolve the problems associated with split-estate energy development, landowners sought resolution to their problems with energy developers via the intervention of state legislators. The conflict between energy developers and ranchers heightened because, in seeking the state's intervention, ranchers were requesting that the state enact legislation that oftentimes superseded federal law and regulation of energy development activities. Their legislative requests for extended notification, explicit accommodation and compensatory damages, increased bonding requirements, and, in particular, the request for compensating for loss of property value did not sit well with the energy industry. As these issues were debated, energy and ranching organizations lobbied their state-elected leaders. In turn, the conflict and competition between the two interests intensified.

To some, the mutual trust that had long benefitted the energy-ranching alliance was shattered. As Bob Gallagher notes, the situation in New Mexico intensified when the first Surface Owner Protection Act was introduced. Gallagher argues that, "We [industry] felt that the trust was lost the very first year when the bill all of a sudden shows up and there hadn't been any conversation about it" (B. Gallagher, personal communication, May 21, 2009). Gallagher also argues that as the debate moved forward, mistrust heightened. Gallagher notes that "Then, the second year, it's okay you came to the table, but there's no trust" (B. Gallagher, personal communication, May 21, 2009). Mistrust, according to Gallagher, led both sides to push each other into corners neither wanted, but because the stakes were so high, the emotional response of each interest group was to further entrench itself into opposing camps. According to Gallagher:

At the end of the day, when you get pushed, at some point you're going to feel like you're in a corner, and what you're going to do then is you're going to react like a caged animal or whatever, and that's when the tensions and emotions got to the point where it was terrible. I mean there wasn't any conversation going on, and the problem is when the main people aren't talking, then what you got here is all these people, these wannabes getting in here and putting other things in here [legislation] that really shouldn't have been in the conversation. (B. Gallagher, personal communication, May 21, 2009)

This view is echoed by Stan Dempsey as he notes that negotiations among stakeholders grew tense during discussions of the first Surface Owner Protection Act introduced in Colorado. According to Dempsey, "During the Curry [Rep. Curry] bill, there were private negotiations outside the capital between some parts of industry and the homebuilders, and they worked for many months to try and come up with something and it was all done quietly. She [Rep. Curry] agreed to carry a bill for the environmental community and the homebuilders, well, that's the first caution, to carry a bill for [them]. And then, she was trying to work with industry. So she was torn apart three different ways. Um, we had some pretty harsh words" (S. Dempsey, personal communication, March 24, 2009). Bruce Hinchey notes that things grew "bitter" as the conflict over split-estate energy development reached the Wyoming legislature (B. Hinchey, personal communication, June 12, 2009). Hinchey comments, "I think in the end, as the bill worked forward there was a lot of emotion initially, of course, with some of those folks [ranchers]. We [energy industry] thought it [law and regulation] was fine the way it was" (B. Hinchey, personal communication, June 12, 2009). State legislative intervention was, in the opinion of industry representatives, unwarranted. Surface Owner Protection Acts, according to representatives of the energy industry, were in response to a minority of landowners whose experience with split-estate energy development had become over-sensationalized.

The energy industry opposed enactment of state Surface Owner Protection Acts. Industry was opposed because it believed that Surface Owner Protection Acts would impose new, unnecessary requirements on industry's ability to develop split-estate energy resources. From the perspective of the energy industry, Surface Owner Protection Acts that sought to impose new requirements of industry were overly burdensome. In the opinion of industry representatives, the Acts are disproportionate responses to the few ranchers and landowners whose problems with energy activities have been blown out of proportion. As Kathleen Sgamma notes, "In the cases where they [energy developer and rancher] can't some to an agreement, those are the stories that get blown up. It doesn't take many people making a fuss to catch the media's attention. It only takes one or two [ranchers] to organize and raise a huge fuss" (K. Sgamma, personal communication, March 24, 2009). Bruce Hinchey adds, "Yeah that's what was reported, just like anything, you've got a handful of people doing all the complaining. There's a handful that are not [happy] and those were the ones doing the complaining. And they're the ones that made all the press and that's what you read about. And, it's just like any law. Any law that gets passed you got a handful of people that are complaining and usually that's the way it goes. They're the ones that's gonna' get the wheel greased if you get something passed" (B. Hinchey, personal communication, June 12, 2009). In a similar vein, Bob

Gallagher refers to Surface Owner Protection Acts as "Bad Actor Acts" (B. Gallagher, personal communication, May 21, 2009). Gallagher notes, "I think we need to rename it the Bad Actor Act because it surely is for bad actors. It's for the oil and gas companies that want to run roughshod over you [rancher], or it's for the rancher at the gate who says 'I don't care what you say or what you're offering, you're not coming on [to the land]" (B. Gallagher, personal communication, May 21, 2009).

The energy industry opposed any legislation that would give split-estate property owners the right to veto an energy company's access to the mineral estate. According to Bob Gallagher, "I buy into that the mineral must be allowed to be produced. I think the bottom line is you just can't be denied. Denied or delayed, and that was our philosophy from the very start when we went into any negotiations. If you have a partnership and it's 50/50 and it's two guys, what happens when it's a tie vote" (B. Gallagher, personal communication, May 21, 2009)? The federal mineral estate and the dominance it carries "covers a lot of ranchland" (K. Sgamma, personal communication, March 24, 2009). Kathleen Sgamma notes that the primacy of the mineral estate should remain absolute (K. Sgamma, personal communication, March 24, 2009). She believes that if the estate were on equal legal footing, the access of energy companies would be denied. Sgamma argues that, "It's problematic if that primacy is taken away because if you don't have that primacy of the mineral estate, [what rancher] is going to say, 'Oh, yeah, come on to my land. I don't own the minerals. I'm not going to get anything out of this except maybe some surface damage money'" (K. Sgamma, personal communication, March 24, 2009). Additionally, the energy industry opposed legislation that would have allowed compensation for the loss of real estate value or its potential for development value. As Bruce Hinchey notes, "When you have landowners that want, for example, land that's worth $100 an acre and he wants $25,000 an acre, then that's a little absorbanent [sic]. And part of it was they [ranchers] also talked about at one point, well I might want to build a hotel out here in the middle of nowhere. I might want to build an amusement park. Of course those are exaggerations, they wouldn't do that, but those are ways to say the land could be worth way more than what you're currently valuing it for" (B. Hinchey, personal communication, June 12, 2009). Finally, the energy industry opposed legislation that would have increased bonding requirements and fees. Bruce Hinchey offers the example of "a guy out of Pavilion [Wyoming] that wanted basically his entire ranch value" (B. Hinchey, personal communication, June 12, 2009). Hinchey adds that the request for the ranch's value as bond was that "he was asking for, what was being offered is then, his ranch was for sale, and I thought, 'geez I could buy his ranch and in 20 years pay the whole thing off with just the lease agreement. Things like that go on" (B. Hinchey, personal communication, June 12, 2009).

Industry representatives are united in their opposition of the surface owner accommodation issues listed above. Initial proposals for Surface Owner Protection laws, in one form or another, included these elements. And because of this, the energy industry opposed their enactment. This is because the energy industry viewpoint is that unlimited access to develop the federal mineral estate is paramount to their economic interests. Additionally, industry representatives view compensation for real estate value or its potential value as overly speculative and, therefore, too expensive a burden for their industry to bear. The same economic viewpoint applies to proposed increases to bonding requirements and fees. Increasing bonding requirements and fees, in the shared opinion of industry representatives, would make development of energy resources too costly. In turn, Surface Owner Protection Acts proposing to implement such measures were resisted by the energy industry. Measures such as these would accommodate split-estate property surface owners, but because they would impose economic burdens on energy companies, the energy industry sought to prevent their inclusion in the Surface Owner Protection Acts initially proposed by ranching organizations. The differing views regarding the level of accommodating surface owners and the cost of those accommodations are at the heart of the contention between energy and ranching over enactment of Surface Owner Protection Acts.

Issues of access, compensation, and bonding were stumbling blocks to the passage of Surface Owner Protection Acts. Bruce Hinchey argues that state legislative intervention to accommodate a few ranchers doesn't make sense. Hinchey argues, "Well, that mineral belongs to me and every other citizen of that state and you're gonna' give something to some rancher that bought some land that didn't own the minerals in the place? That doesn't make sense. You're [government] taking away. They bought the land knowing they weren't going to get minerals" (B. Hinchey, personal communication, June 12, 2009). Issues of accommodation contributed to the general lack of communication between the energy and ranching industries. In turn, debates over access, compensation, and bonding heightened the competition for control of land-use policymaking. Simply stated, these legislative debates were proxy wars as each interest sought control of the BLM's land-use subgovernment.

Annexing the BLM's Land-Use Subgovernment: Energy's Perspective

Energy industry officials have varied beliefs regarding whether their industry has taken the BLM's land-use subgovernment away from ranching. Opinions regarding the ranching industry's deference to energy development interests within the BLM's land-use subgovernment were nuanced. In the opinion of industry representative, the answer to the question of whether or not a shift in control of BLM land-use decision-making favors the energy industry is unclear. Many believe that if ranchers have a dispute, the dispute is not with energy as much as it is with the BLM. In Bruce Hinchey's opinion:

> I always thought there's been equal footing. And I thought ranchers got along quite well with industry. 'Least the vast majority. I still think that's the case and think there's mutual respect because it's their land. It's not our land. And we're gonna' be there for a short period of time to use it. And then when we're gonna' be gone. They'll still have their land. We go back in and we reclaim it [the land], and in a lot of cases what is interesting is the ranchers are mad at the BLM. Because when we go in to reclaim we have to reclaim to their [BLM] standards and not the standards of the ranchers. (B. Hinchey, personal communication, June 12, 2009)

Bob Gallagher also notes that ranchers' complaints should rest with BLM decision-makers regarding land-use conflicts. Gallagher argues that, "If you're a landowner, and if you [BLM] gave a permit to that guy to graze on your [federal] land, then if that guy's [rancher] got a problem, he ought to come to you [BLM]. If the BLM gave a permit to graze and we cut a road through there and we produced and all of a sudden they can't have a thousand head grazing, they can only have eight hundred, well, they have a legitimate concern. Why is the beef with us? We have the same type of permit to be out there as they do. Their beef needs to be with the BLM, but yet they want it to be with us" (B. Gallagher, personal communication, May 21, 2009).

The energy industry's dominance of the BLM's land-use decision-making is limited by the BLM's bureaucratic entrenchment. Gallagher argues that, "For someone just to say blanket-wise, 'oil and gas dominate that, and they [BLM] really don't care over here [grazing],' I agree somewhat in part, but it hasn't gone full-circle, and I think that the reason it hasn't gone full circle is because of what I call the 'BBs,' and those are the bureaucrats, and those are the guys that when the political appointees come in, the guy sits behind his desk and the bureaucrat will say 'I'll be here when you get here and I'll be here when you gone'" (B. Gallagher, personal communication, May 21, 2009). Gallagher concludes that, "They [BLM bureaucrats] still have that thought process that the BLM was really created for range and wildlife and this and that. It's [domination] shifted, but I don't think to the point where it's [grazing and wildlife] totally ignored. When you look at it now we're [energy industry] not even waiting 30 days on permits [APDs]. How did that happen? There's more money put into that, there's no doubt about it" (B. Gallagher, personal communication, May 21, 2009). Additionally, Kathleen Sgamma regards the conclusions that "the relative rise or fall of ranching power, vis-à-vis, the BLM" is a conclusion that must be tempered through the lens of the BLM's bureaucratic decision-makers (K. Sgamma, personal communication, March 24, 2009). Sgamma concludes that, "I think BLM obviously has a tough job. I mean, they're never going to satisfy everybody. They're getting beat up by environmentalists for whatever they do, allow ranching, allow oil and gas, any other mineral development, so there's that constituency that is always pounding on them to do nothing with the land" (K. Sgamma, personal communication, March 24, 2009). Thus, representatives of the energy industry do not agree with the presumptive conclusion that their industry has come to dominate the BLM's land-use subgovernment. As representatives of their industry's interests, they believe that control of the decision-making process is tempered by the democratic engagement of the multiple interests at stake in governmental decisions regarding land-use.

If change or reform is desired, the democratic decision-making process will reflect the public's desire for alteration of industry primacy in land-use. From the perspective of the energy industry, BLM decision-makers are following the law and existing rules and regulations. As Kathleen Sgamma notes, "You talk to the oil and gas people at the BLM, and they're just following the law; they're doing what the law says now, and getting criticized by all quarters for it. But, if you don't like the law, there's a democratic process in place to change it" (K. Sgamma, personal communication, March 24, 2009).

Split-estate landowners used the democratic process in seeking to protect their interests from harm, but change or reform to existing law was met with resistance by the energy industry. Simply stated, debates over states enacting Surface Owner Protection Acts were democracy in action. However, state legislation sought to reform existing federal laws and regulations guiding split-estate energy development. As Bruce Hinchey suggests, "Well, I think it's the same as it was. I never saw a change from the time before the bill was passed, after the bill was passed, till now. There hasn't been, nothing's changed as far as I'm concerned" (B. Hinchey, personal communication, June 12, 2009). Therefore, the impact of the states' legislative efforts to protect split-estate property owners—primarily ranching or agricultural interests—from the harm of energy development was limited on the one hand by federal law and, on the other hand, through the lobbying efforts of the energy industry. As Bob Gallagher notes:

> There were two or three years in a row that they [ranchers] attempted to pass a Service Owner Protection Act in New Mexico and not include oil and gas around the table when they did it, and we [industry] had to kill it. We killed it two years in a row. All of a sudden, the next year it was obvious to us that something was going to happen, and I love to say, 'if you're getting ready to get thrown out of town, get in front and make it look like a parade.' (B. Gallagher, personal communication, May 21, 2009)

Ranchers and energy developers have a history of resolving their differences amicably without the intervention of either state or federal governments. As these organizations have interacted within the BLM's land-use subgovernment for decades, tradition holds that a rancher's request for resolution of a problem with an energy developer is handled by a representative of the energy company quickly and quietly. The expansion of split-estate energy development, however, altered the traditional manner in which ranchers and energy developers resolved their differences.

Conclusion: Dominance of a Subgovernment

The expansion of domestic energy development encroached upon the privately owned split-estate surface lands of ranchers and homeowners across the West. As split-estate energy development expanded, conflicts between energy and ranching interests multiplied. The conflict was particularly problematic in areas where development of non-traditional energy resources, such as CBM, spiraled to record numbers. As drilling increased, problems increased, and with them, complaints among surface-owning landowners increased. As a result, ranchers began taking their complaints to the BLM. As noted by BLM administrators, the agency is confined by law and regulation regarding domestic energy development in its capacity to resolve the conflicts that may occur between a surface owner and an energy developer. Simply stated, the BLM is not in the business of conflict resolution.

As the energy industry took advantage of the favorable conditions for energy development, the complaints of split-estate landowners gained the attention of state lawmakers. Split-estate landowners sought the attention of their state lawmakers because the BLM is restrained in its ability to address and resolve conflicts between energy developers and landowners. As complaints over split-estate energy development reached state lawmakers, the conflict between energy developers and ranchers heightened. Among those complaining loudest were ranchers who believed that the transgressions of energy developers were not being adequately addressed by either the energy development industry or the BLM. In turn, organized groups of ranchers and landowners petitioned their state-elected officials for protection from the unlimited access of energy developers and the harm that split-estate energy development was creating. Motivated by constituent complaints, state lawmakers across the Rocky Mountain West introduced bills known as Surface Owner Protection Acts.

Surface Owner Protection Acts further heightened the conflict between energy and ranching. Unlike their past experiences with government intervention, energy and ranching interests were no longer aligned in protecting the legal and regulatory status quo regarding their shared land-use. Instead, energy development and ranching interests were now competing with each other as each interest sought to protect itself from the harm of the other interest. On the one hand, ranching and landowner organizations sought to protect themselves from harm by advocating reform of the legal and regulatory status quo concerning split-estate energy development. On the other hand, energy development organizations sought to defend the legal and regulatory status quo of split-estate energy development from being reformed.

As energy and ranching organizations disagreed over state legislation, they interacted with each other in an increasingly hostile manner. As a result, the traditional alliance of energy developers and ranchers became increasingly strained. This is because each interest recognized the stakes in securing a successful outcome from state lawmakers were high. As their legislative battles unfolded across the Rocky Mountain States of New Mexico, Colorado, and Wyoming, energy developers and ranchers sought the same end: dominance of the BLM's land-use subgovernment.

References

Dempsey, S. (2009). President of the Colorado Petroleum Association (CPA). Interview Conducted: March 24, 2009; Denver, CO.

Gallagher, B. (2009). President of the New Mexico Oil and Gas Association (NMOGA). Interview Conducted: May 21, 2009; Albuquerque, NM.

Hinchey, B. (2009). President of the Petroleum Association of Wyoming (PAW) and Former Speaker of the House, State of Wyoming Legislature. Interview Conducted: June 12, 2009; Casper, WY.

Sgamma, K. (2009). Director of Government Affairs for the Independent Petroleum Association of Mountain States (IPAMS). Interview Conducted: March 24, 2009; Denver, CO.

Chapter 7
Ranching

Abstract Ranchers are not unified in their hostility toward energy developers. Among ranchers, there are those who favored the legal and regulatory status quo of split-estate energy development and those who favored reforming current policies. Among ranchers favoring the status quo are those who believe that legislative intervention would negate their ability to negotiate terms with energy developers seeking to access and develop the federal mineral estate. Among ranchers favoring reform, most believe that legislative intervention would enhance their ability to negotiate terms with energy developers. In either case, the central focus of State Surface Owner Protection Acts was reformation of the negotiation process wherein ranchers entering into contractually binding Surface Owner Agreements with energy companies. Representatives of ranching organizations express the opinion that Surface Owner Protection Acts are beneficial to all their members. Unlike energy representatives, however, representatives of ranching are uniform in their opinion of how the ranching-energy conflict heightened.

Keywords Ranching lobby · Bureau of Land Management · Department of Interior · Bureaucracy · Split-estate energy development · Surface Owner Protection Act

The Voice of Ranching

The BLM has shifted from a rancher-dominated agency to an energy-dominated agency. The expansion of domestic energy development across the states of the Rocky Mountain West, coupled with an increasing demand for energy, caused the BLM to shift its policy emphasis and resources away from grazing and ranching to energy development. As energy development expanded and BLM energy policies and resources shifted, ranching operations across the West were negatively impacted. These negative impacts are particularly acute among ranchers whose operations are

---

"If you take the microcosm of Wyoming, and you talk about the Bureau of Land Management since the late '90s, if you're going to dub it anything, it's going to be the Bureau of Energy Development." (Jim Magagna, Director of the Wyoming Stock Growers Association)

© Springer Nature Switzerland AG 2019
R. E. Forbis Jr., *Altered Policy Landscapes*,
https://doi.org/10.1007/978-3-030-04774-0_7

127

located on split-estates. As energy developers sought to develop the federal mineral estates underneath the surfaces of ranching operations, split-estate energy development triggered conflict between ranchers and energy developers. In turn, ranching organizations sought to protect their interests by petitioning their state legislatures to enact Surface Owner Protection Acts.

Surface Owner Protection Acts, clearly opposed by energy development organizations, were favored by most, but not all, ranching organizations. As deliberation of state legislation took place, factions of ranchers emerged, who, like their energy development brethren, opposed enactment of Surface Owner Protection Acts. It became clear that among ranchers there are differing opinions regarding the extent to which ranching's interests were negatively affected by the expansion of domestic energy development.

Ranchers are not unified in their hostility toward energy developers. Among ranchers, there are those who favored the legal and regulatory status quo of split-estate energy development and those who favored reforming current policies. Among ranchers favoring the status quo are those who believe that legislative intervention would negate their ability to negotiate terms with energy developers seeking to access and develop the federal mineral estate. Among ranchers favoring reform, most believe that legislative intervention would enhance their ability to negotiate terms with energy developers. In either case, the central focus of State Surface Owner Protection Acts was reformation of the negotiation process wherein ranchers entering into contractually binding Surface Owner Agreements with energy companies.

Representatives of ranching organizations[1] express the opinion that Surface Owner Protection Acts are beneficial to all of their members. Unlike energy representatives, however, representatives of ranching are uniform in their opinion of how the ranching-energy conflict heightened. Representatives of ranching believe that energy markets and technological advancements gave the energy industry its capacity to develop non-traditional energy resources. They also believe that the Bush Administration's exercise of wielding executive power greatly influenced the expansion of energy development. Additionally, representatives of ranching share the opinion that the behavior of energy developers toward split-estate property owners and the damages that resulted from energy development are primary causes for heightening their conflict and competition with the energy industry. They also share the belief that the energy industry's opposition of to Surface Owner Protection Acts negatively impacted the historical solidarity of the ranching-energy alliance.

Ranchers' land-use interests remain intertwined with those of the energy industry. Environmental interest groups, primarily conservation-oriented interest groups,

---

[1] Note: Ranching participants include:

1. Jim Magagna, Director of the Wyoming Stock Growers Association (WSGA)
2. Caren Cowan, Director of the New Mexico Cattle Growers Association (NMCGA)
3. John Vincent, Legal Counsel to the Landowners Association of Wyoming (LAW), Mayor of Riverton, WY
4. L. Goodman, Chief Legislative Lobbyist for the Landowners' Association of Wyoming (LAW)

were a factor in ranching's legislative battles with the energy industry. Ranching's alignment with environmental organizations was a source of tension as ranchers and energy developers tangled over the enactment of Surface Owner Protection Acts. Representatives of ranching express that their alignment with conservation-minded environmental organizations will continue to develop. However, they also express the conviction that ranching's continued alignment with energy developers is in the best interest of ranchers. Finally, representatives of ranching organizations are united in their belief that, due to a variety of factors—not the least of which is the expansion of domestic energy development—ranching's former dominance of the BLM's land-use subgovernment has been annexed by the energy industry.

The voices represented here are those of ranching elites. They represent the perspective of both traditional ranching organizations and the splinter groups that emerged from those organizations as each of these groups of ranchers engaged in the legislative battles over Surface Owner Protection Acts. These splinter organizations are known simply as "landowner" organizations. They are defined as elites because, as stated previously, they are a select sample of actors who interact routinely with other interest groups and governmental entities that compose the networks of the BLM's land-use subgovernment. These elite actors were, at the height of ranching's conflict and competition with the energy industry, active participant representatives of ranching's interests.

Expanded Energy Development Disrupts Ranching Operations

Split-estate energy development triggered conflict and competition between ranchers and energy developers. As domestic energy development activities on split-estate ranchlands increased, ranchers became dismayed with the behavior of energy development companies toward landowners. Ranchers also grew increasingly angry over damages to their ranchlands that were the result of increased levels of energy development activities. Caren Cowan, Director of the New Mexico Cattle Growers Association, remarks that, "Our folks have been concerned about the damages of oil and gas production on their land for probably 40 years. But, over the last 10 or 12 years as the price of energy has increased, and there's been more pressure put on the land to produce more energy, it's become a bigger and bigger issue for members across the state" (C. Cowan, personal communication, May 21, 2010). Ranching's anger with energy developers heightened as problems associated with split-estate energy development went unresolved. Jim Magagna, Director of the Wyoming Stock Growers Association, adds, "That's what caused me to get more and more calls from my members saying, 'You know, we are willing to work with these companies but there are some things happening out there that are just in total disregard to our interests, and they need to be addressed" (J. Magagna, personal communication, March 23, 2009). Therefore, as a result of energy's growth and its inattentiveness to ranchers' concerns, the trust that had been the hallmark of the alliance between ranchers and energy developers began to deteriorate.

Rancher mistrust of energy developers increased as split-estate energy development activities increased. In particular, ranchers grew increasingly suspicious over the earnest nature of energy developers in their negotiation of Surface Owner Agreements. As Jim Magagna notes, suspicion was often the result of a company's unfamiliarity with ranchers and their culture. Magagna comments that:

> As Wyoming began to experience rapid growth in mineral [energy] development in the late '90s and early 2000s, and particularly as we [Wyoming] moved away from just the traditional large mineral operators in the state, you know, the companies that had been here and done business every year and had a presence, we began to have more and more new companies coming in with coalbed methane [CBM] development. We saw a lot of what have sometimes been termed fly-by-night companies. I think some of them fit that description well; some were very reputable companies, but smaller operators who had not operated in Wyoming before, and frankly, had not operated in an area where split-estate is so dominant. (J. Magagna, personal communication, March 23, 2009)

In addition to new energy operators' unfamiliarity with ranching culture, another source of tension was the subdivision and sale of once-large tracts of ranchlands created tension within the ranching community. Most property sales conveyed only the surface rights; thus, there are more split-estate property owners today than there were during previous periods of increased energy development. Caren Cowan comments that "part of the complexion of things [is that] the land is getting split up more and more." She notes that the parceling of large ranch acreage and selling of those parcels for their real estate value have occurred "probably more in Colorado than here [New Mexico], but probably more here than in Wyoming" (C. Cowan, personal communication, May 21, 2010). The shift in Western landownership resulted in the previously discussed increased "urban interface." Urban interface is cause for conflict because persons migrating to the West, hopeful of owning a bit of Western tranquility, were instead confronted with the fast-paced reality of energy development.

Energy developers unfamiliar with the culture of ranching are interacting with landowners who are just as unfamiliar with culture of the energy industry. One cause of landowner unfamiliarity stems from their presumption that it is unlikely their property will be developed for its energy. Cowan adds that "somebody would buy 5 or 10 acres, build their dream house on it, and then all of a sudden one day have somebody [land-man] knock on the door that says 'we're going to take your two back acres or three acres for an oil pad, and we're going to build a road right here beside your house, and we're going to pay you $5,000 for this,' or you know, whatever, and we [Cattle Growers Association] heard a lot of those kind of stories" (C. Cowan, personal communication, May 21, 2010). Landowners such as those described by Cowan are residents of a New West. Cowan reasons that part of what has shaped the ranching-energy conflict is that "people come out here [West]; they don't understand mineral estate and dominance and all that stuff. They bought a piece of property; they didn't know enough to see who owned the mineral under what they're buying, so all of a sudden, they're extremely upset" (C. Cowan, personal communication, May 21, 2010).

New West landowners' unfamiliarity with their property rights is a common theme among representatives of both ranching and energy organizations. There is, however, a difference of opinion regarding who is ultimately responsible for knowledge and understanding of the mineral estate's legal dominance over the surface estate. As discussed in the previous chapter, representatives of the energy industry believe that landowners are responsible for knowing who controls the mineral estate. Representatives of ranching organizations believe that information regarding who controls the mineral estate should be made clear when prospective buyers are considering the purchase of property. Unlike their energy counterparts, representatives of ranching suggest that there exists the possibility that real-estate brokers underemphasize the possibility of energy development occurring on or within proximity of a property being sold. Potential landowners presume that because they are purchasing such small amounts of acreage, it is implausible that energy development will impact their property. To which Cowan responds, "Yeah, Wrong" (C. Cowan, personal communication, May 21, 2010)!

When energy development occurs, the intensity of the activity is often overwhelming to landowners. John Vincent, legal counsel to the Landowners Association of Wyoming, observes that "what happened is that technology changed, so instead of having one gas well per 640 acres [the original homestead acreage], it [well-spacing] went to 160 acres and then to 40 acres and then the last rule change I think is 1 well to at least every 20 acres, and it may now be down to 5 or 10 [acres per well]" (J. Vincent, personal communication, March 16, 2009). Vincent argues that "the development not only became more intense, just in terms of the number of wells allowed per section, it became more intense in the sense that you didn't necessarily have to drill them all on centers [center of the acre from the section-line]. So you could have an area [of surface property] that was just flooded with wells" (J. Vincent, personal communication, March 16, 2009). He uses the example of "wells piling on" top of a surface property to such a degree that the property itself became unusable as the owners had intended (J. Vincent, personal communication, March 16, 2009). Vincent describes that "[overlay documents illustrating changes to the property over time] show the ranch when they [the landowners] first bought it. Actually, this was a big hay farm that they used in conjunction with the ranch, but it had maybe four wells. We [law firm] had overlays that just showed how these wells kept piling on and basically it ruined the farm as a farm. It was such that you couldn't irrigate it, you couldn't run equipment on it and, [as a result] they no longer own the farm" (J. Vincent, personal communication, March 16, 2009).

The legal dominance of the mineral estate development comes at the expense of the property owner's development of the surface estate. Unless the owners of the surface estate understand the legal preference of the mineral estate's use and the regulations that guide that use, they are not in a position to deter access. Unable to deny access, landowners are limited in their ability to mitigate the harm that may result from energy activity. As is required by law, access to the mineral estate is negotiated between the landowner and the energy developer's representative. As part of the use agreement, compensation for foreseeable disruption of the surface estate and the harm that may result from energy activities are negotiated with energy development companies.

Split-Estate Surface Owner Agreements

Split-estate property owners lack equal footing in their ability to negotiate terms of access and compensation. The legal dominance of the mineral estate over the surface estate influences the bargaining position of split-estate property owners. Ranchers acknowledge the legality of the mineral estate's dominance over the surface estate. Ranchers, however, do not like the term "dominance" (Interviews collectively). Representatives of ranching organizations contend that their conflict with the energy industry stems from the mineral estate's legal dominance. In turn, they regard the concept that ranching's development of the surface estate as being subservient to development of the federally owned mineral estate as fundamentally unfair. They argue that when the conditions of political willpower, energy markets, and technological advancement aligned in a manner establishing energy development as the preferred use of the land, any power ranchers might have once held in their ability to negotiate with energy developers was diminished.

President Bush's unilateral use of executive power disrupted the equal footing of ranchers and energy developers. The Bush Administration's ability to shift the energy policies of the BLM affected the ability of landowners to conduct ranching or agricultural operations. In ranching's parlance, the effect of the administration's meddling with the BLM's energy policy and the subsequent expansion of domestic energy development affected surface use to such a degree that it became increasingly difficult to "run cattle." Caren Cowan remarks that, "we were very concerned and disappointed that equal consideration wasn't given to the surface when those executive orders came out because that just sort of turned everything loose. You began to see such concentration on energy production that we can't get things done on grazing allotments. Everybody [BLM] is tied up doing whatever they had to do to get the next energy project going, and us [ranchers] who were having problems on the grazing end of it are just left hanging out to dry" (C. Cowan, personal communication, May 21, 2009). Cowan confirms that as a result of the Bush-Cheney emphasis on expanding domestic energy development, ranching organizations witnessed a shift in BLM "personnel and budget" (C. Cowan, personal communication, May 21, 2009). Jim Magagna echoes Ms. Cowan's observations as he argues that a shift in BLM policy and resources were the result of multiple factors. Magagna concludes that:

> There was the Bush administration that was friendly to mineral development, and that certainly fostered it [shift in BLM policy and resources]. I wouldn't disagree, but I think that eliminates a number of steps in between. Certainly the Bush action led to the intensity of the desire to develop. But other things have played in there that were very important. Well, technology. You look at the massive Jonah Field in Western Wyoming. Twenty years ago there was not technology to produce that. You look at all the coalbed methane development in the Powder River Basin [Northeast Wyoming]. Twenty years ago no one envisioned coalbed methane as being a source of natural gas. The technological changes and the political atmosphere sort of came together as a point in time, I think early 2000s, late '90s, and chicken and egg thing. I don't know which came before the other, but I think that what happened would not have happened without the concurrence of both. (J. Magagna, personal communication, March 23, 2009)

Finally, John Vincent adds that with the 2000 election of the Bush Administration "we, of course, saw the energy development really ramp up. And what we saw at that time around here locally [Wyoming] was the development of natural gas fields" (J. Vincent, personal communication, March 16, 2009). The expansion of energy, therefore, caused a shift in BLM policy and resources, redistributing the BLM's balanced approach to land-use. In turn, there was a redistribution of negotiating power between ranchers and energy developers.

The inequity in negotiating power is compounded by energy development's expansion onto split-estate properties. The inequity of negotiating power between ranchers and energy developers is reflected in the difficulty split-estate property owners have in negotiating the terms of access and accommodation with energy developers. One reason for the disparity in negotiating power is the economic power of energy. The price of a barrel of oil is no longer equal to a pound of beef on the hoof (Interviews collectively). In market terms, this directly affects a rancher's ability to negotiate terms of a Surface Owner Agreement. Therefore, the legal, political, and economic equity ranchers once enjoyed in their ability to negotiate with energy developers is greatly diminished. As Jim Magagna summarizes:

> In my mind when you're talking about split-estate you've got two property rights. One is as valid as the other, but I also accept that it's pretty well established, certainly in Wyoming, that the mineral estate is a dominant estate. Now, I hate that terminology because I think it's asking for a fight when you say, 'I dominate you.' But the reality is that properly interpreted, there is the dominant estate, which means an absolute right to come on your property as needed in order to produce the mineral. So in that sense you'd like to think that it's two property rights. The reality is you've got two property interests: one is dominant; one is also more powerful, and more knowledgeable. I mean, knowledge is power, and part of it is not that they're [energy developer] ruthless or that they are bad people, it's that they may be doing thousands of these [Surface Owner Agreements], and here I am as a landowner or rancher doing my first one or one of a handful. So, sure, are there mineral companies that are just ruthless? Yes. Are there ranchers that are equally ruthless? Yes. And so, if you are going to just say I don't want to get along, it's just as easy for one side or the other to say that. But when you say I do want to get along and I want to negotiate, then obviously the party with more experience, more resources can be a little more heavy handed in the negotiations. (J. Magagna, personal communication, March 23, 2009)

The party with the most legal, political, and economic power has the upper hand in negotiating the terms of access and accommodation in Surface Owner Agreements. Having greater knowledge and information is also beneficial in controlling the course of the negotiations concerning the manner in which access will occur or the type of accommodations being made. Often, as Caren Cowan notes, surface owner's lack of knowledge and information is the cause for friction between a rancher and developer. As Cowan remarks, "Yeah, somebody enters into an agreement and then calls after the fact and says, 'what should've I asked for?' You know we all are busy, we all think we can handle our own business, and we don't ask for help early enough often enough. So that's a continuing problem, and people don't share information with each other" (C. Cowan, personal communication, May 21, 2010). One reason why information is not communicated is that the negotiation process and the deal struck from those negotiations are private.

Problems associated with surface owners' lacking knowledge and information arise in part because the conditions of Surface Owner Agreements are not standardized. As discussed, the reason Surface Owner Agreements are not standardized is that conditions of the contractual agreement remain unregulated by the government. This means that issues of access and compensation for any foreseeable surface disruption and/or damage are privately negotiated between the individual parties. In cases where split-estate landowners cannot afford competent legal counsel to represent their interests, they are at the mercy of the well-armed energy company's representative. As is often the case in negotiations such as these, split-estate landowners are left to make the "best deal they can get" (Interviews collectively).

The best deals are often struck between ranchers and energy development representatives most familiar with the expectations of each side in the negotiation of Surface Owner Agreements. Jim Magagna adds that, "A lot depends on who the land-man is and who the rancher is" (J. Magagna, personal communication, March 23, 2009). Generally speaking, representatives of the ranching industry concede that conflict can be avoided if land-men were to approach landowners in a manner that highlights cooperation rather than confrontation. Magagna adds that the approach taken by land-men within the same company can differ. He remarks that, "I've had a rancher tell me that the last land-man that company had was [the] worst company I've ever dealt with, and the current land-man they have, that's the best company I've ever dealt with. So, it [conflict or agreement] depends on the approach" (J. Magagna, personal communication, March 23, 2009). Magagna's remarks underscore ranchers' frustration with the legal dominance of the mineral estate. Caren Cowan argues that each side's lack of familiarity with what is expected during the course of negotiating Surface Owner Agreements stems from the frustration of ranchers over the issue of mineral estate dominance. Cowan adds that frustration with the legal dominance of the mineral estate, in part, is related to the approach taken by BLM administrators in defense of the mineral estate's development. She remarks that, "We [ranchers] understand that that's federal law. It's frustrating that it appears that the BLM thinks that an oil and gas lease is a right where grazing is a privilege. We don't see it that way, obviously, but the BLM does" (C. Cowan, personal communication, May 21, 2009). The implication is that the privilege of grazing one's livestock is deferential to the right of access and development of the oil and gas lease. Therefore, because the BLM views oil and gas leases as having a property interest, the surface owner cannot legally deny an energy company's right of access and development of the mineral estate. According to John Vincent, the inability to deny access and development places landowners in a precarious bargaining position (J. Vincent, personal communication, March 16, 2009). Vincent notes that, "The predicament of ranchers and farmers [is] that they just absolutely don't have a bargaining position; even enough of a position to insist that things like, 'if you can directionally drill a well from the side of a hay field, do that rather than plopping a well right in the middle of the field'" (J. Vincent, personal communication, March 16, 2009).

The BLM's defense of the energy lease as a property right is reflected in the ability of its allowing energy companies to post a bond if a Surface Owner Agreement cannot be reached.[2] Posting a bond guarantees an energy company's right to access the mineral lease. As noted previously, all that is required of an energy company to develop the energy resource is the purchase of a federal mineral lease, an APD, and provide notice that a Surface Owner Agreement has been reached with the landowner. The assumption of BLM administrators is that within the Surface Owner Agreement contract, sufficient accommodation and compensation is made to the landowner. The presumption that landowners are sufficiently compensated for the energy industry's use of their surface lands harkens back to days when ranchers and energy developers reached handshake agreements.

A Handshake Deal Is Not What It Used to Be

As split-estate energy development intensified, problems associated with its development multiplied and ranchers sought to protect themselves from the harms of the energy activities. In particular, ranchers were growing increasingly concerned over the unfettered access and development by energy companies on their surface lands. Sensing that greater harms were forthcoming if they did not begin to address the inherent inequity of the law, ranchers sought the protection of their state legislators. As ranchers sought to defend their interests through the legislative process, their relationship with the energy industry worsened. Each interest began to compete for control of land-use policies relative to split-estate energy development. As each of the interests sought to defend its position to state legislators, the conflict and competition between ranchers and energy developers heightened. Simply put, viewing the conflict and competition through the lawmaking process, it is clear that the historical alliance of ranching and energy unraveled.

Mitigating the impact of the mineral estate's legal dominance over the surface estate was of primary legislative importance to ranchers. As mentioned previously, ranchers do not like the term "dominance," but energy developers came to rely on the mineral estate's continued legal dominance to develop energy resources. Jim Magagna regards the mineral estate's dominance as a "boiler-plate" issue (J. Magagna, personal communication, March 23, 2009). Magagna explains, "Yeah, [ranchers] don't like the term, but on the other hand, I kid my friends in the mineral industry that we're just going to start putting [a clause] in every bill, it doesn't matter if it's about health care or what it is, that says the mineral estate is dominant because they're almost paranoid about repeating that as often as they can legislatively" (J. Magagna, personal communication, March 23, 2009). The question of legally protecting access and development of the mineral estate is, according to John Vincent, a matter of which party is behaving reasonably. In Vincent's opinion, the legal defense of the mineral estate should be balanced with the legal defense of the surface estate (J. Vincent, personal communication, March 16, 2009). Vincent notes that if defense of the estates were balanced, "Nobody

---

[2] Note: See generally discussion in Chaps. 4 and 5.

would have the upper hand. The question then is, whether the landowner is being unreasonable and refusing to let the oil company do anything, or is the oil company being unreasonable? What that [balance] does is drive people to a position where they have to negotiate fairly because nobody has a whip hand" (J. Vincent, personal communication, March 16, 2009). Vincent argues that "the fact of the matter is that the oil companies have the whip hand" (J. Vincent, personal communication, March 16, 2009). The legal and regulatory dominance of the mineral estate, and the BLM's defense of its dominance as a use of the land, provides energy industry leverage in its negotiations with split-estate landowners.

The leverage to dictate terms to split-estate property owners limits the adequacy of Surface Owner Agreements to address and mitigate surface disturbance and harm. If the conditions of the Surface Owner Agreement are agreed upon, the terms of the agreement are unassailable. Bound by the agreement, should harm result from an energy development activity unaddressed in the contract, energy companies are under no legal obligation to compensate the landowner for that damage. And, because conditions of the negotiated agreement are unregulated, unforeseen damages are often not addressed in the final Surface Owner Agreement. John Vincent uses the example of an oil company's subcontractor denial of a landowner's demand of $500 for the repair of a fence to illustrate the nature of a harm being unaddressed in the Surface Owner Agreement. As a result of the company's denial for compensation, the landowner brought suit against the company. Vincent notes that by his estimation the company spent at least $50,000 in legal fees in "just a good, old-fashioned bloodletting over $500" (J. Vincent, personal communication, March 16, 2009). Agreements over the $500 needed to repair a fence, or similar types of quid-pro-quo arrangements, according to representatives of ranching, were commonplace deals struck between ranchers and energy developers (Interviews collectively). Representatives of ranching organizations lament that the days of sitting down with the land-man and hammering out an equitable agreement over a cup of coffee at the kitchen table are now a thing of the past (Interviews collectively).

Split-Estate Energy Development Reform: Ranching's Perspective

Ranchers experienced in conducting informal negotiations opposed enactment of Surface Owner Protection Acts. Ranchers who were largely unfamiliar with how to conduct negotiations with energy developers favored Surface Owner Protections Acts. As a result, factions of ranchers developed within traditional ranching organizations. These factions, in turn, helped shape the competition between ranching and energy developers. The factions of ranchers who sought to retain the status quo aligned themselves with the lobbying efforts of their traditional ranching organizations. Ranchers favoring reform developed grassroots organizations such as the Landowner Association of Wyoming (LAW). In New Mexico, ranchers favoring reform gravitated between their traditional ranching organization, the New Mexico

Cattle Growers Association (NMCGA), and other grassroots organizations such as the Oil and Gas Accountability Project (OGAP). In all cases, because members of ranching's traditional organizations such as the Wyoming Stock Growers Association (WSGA) and the New Mexico Cattle Growers Association (NMCGA) retained membership in their parent organizations, they were able to influence these organizations to wield their considerable lobbying influence in their respective state legislatures. The influence of traditional ranching organizations assisted in ranching's effort to enact Surface Owner Protection Acts in New Mexico, Wyoming, and, to a lesser extent, Colorado.

Splinter groups of ranchers sought to sever the ties of their traditional organizations with the energy industry. According to Laurie Goodman, Chief Legislative Lobbyist for the Landowners' Association of Wyoming, one of the strategies employed by ranchers favoring reform was to "sever the historical alliance of agribusiness and the energy industry by allowing the voices of their [ranching] own members to articulate their problems" directly to members of the legislature (L. Goodman, personal communication, March 23, 2009). Goodman argues that "empowering individual landowners, members of their [ranching] organizations, to express how the fundamental values of their organizations were not aligning with protecting them from the threats posed by industry [energy], caused them to sever, at least temporarily, from their parent organizations" (L. Goodman, personal communication, March 23, 2009). John Vincent adds that the traditional alliance of ranchers' parent organizations with the energy industry wields tremendous influence on state lawmakers and their decision-making (J. Vincent, personal communication, March 16, 2009). Vincent argues that the collective influence of the traditional organizational alliance of ranching and energy in Western legislatures is "almost as though those lobbyists feel that any legislation that the legislators feel should pass has to be properly vetted with them" (J. Vincent, personal communication, March 16, 2009).

Ranchers favoring reform confronted their parent organizations' entrenched alliance with the energy industry to effect reform. Their direct confrontation created what Ms. Goodman refers to as a "shift in motivating the political scene" (L. Goodman, personal communication, March 23, 2009). Goodman views this shift as the result of a grassroots movement within the parent organizations of ranchers (L. Goodman, personal communication, March 23, 2009). In ranching's legislative battle to protect their interests, Goodman notes that the more knowledge ranchers gained of the energy industry's treatment of split-estate energy development and the harm development activities created for their fellow ranchers, some ranchers began to reassess their alliance to the energy industry (L. Goodman, personal communication, March 23, 2009). As Ms. Goodman observes, "enviros were beginning to be viewed as 'not the enemy' in this battle, but rather, the enemy was now seen as 'one of their own: The oil and gas industry" (L. Goodman, personal communication, March 23, 2009). John Vincent adds that dislodging ranchers' "us versus them" mentality toward environmentalists happened because "you find members that really don't espouse or follow the views of their [parent] organization, and that caused some conflict" (J. Vincent, personal communication, March

16, 2009). Ranchers who did not espouse the traditional views splintered off from their traditional ranching organizations. In doing this ranchers began to form some traditionally unimaginable alliances for the purpose of enacting Surface Owner Protection Acts.

The Intervention of Environmental Organizations

The triumvirate powers of ranching, energy, and environmental organizations that compose the BLM's land-use subgovernment's network of interests were scrambled during the legislative debates of Surface Owner Protection Acts. Prior to their legislative competition, ranchers and energy developers "were still entrenched in the 'us against the enviros argument'" (L. Goodman, personal communication, March 23, 2009). Prior to employing the strategy of shifting the political scenery, ranchers seeking state legislative intervention sought out the respective power brokers of their parent organizations to cut a deal with their fellow energy industry power brokers. As John Vincent and Laurie Goodman both contend, "But those brokers were the ones benefitting the most from the status quo. They were seeking to protect their benefits at the expense of smaller landowners. They were the ones with political access because they had the largest interests invested through land, energy, and mineral ownership" (L. Goodman, personal communication, March 23, 2009; J. Vincent, personal communication, March 16, 2009). Entities of ranchers, however, "began to break free from the influence of these big shadows" and organize themselves into a grassroots movement separate from their parent ranching organization (L. Goodman, personal communication, March 23, 2009). As Ms. Goodman suggests, "We sought to build a grassroots movement through education and empowerment that was unique in that the issue of split-estate energy development was the issue itself. It [split-estate energy development] was the focal point that attracted people with similar values to our cause. We sought to build the parade" (L. Goodman, personal communication, March 23, 2009).

Unlike Wyoming ranchers, ranchers in New Mexico who favored reform remained largely aligned with their parent organization. This is because the New Mexico Cattle Growers, unlike their Wyoming counterpart, took the position of favoring reform efforts early on (C. Cowan, personal communication, May 21, 2010; L. Goodman, personal communication, March 23, 2009). The organization's support of reforming split-estate energy development stems from what Caren Cowan refers to as member frustration with the mindset of the energy industry (C. Cowan, personal communication, May 21, 2010). She argues that members remained largely unified because "It was just some extreme frustration with what was happening to the surface. I mean there are some in the oil and gas industry that just feel the surface is in the way for them to get what they want and need, and what they feel is their right" (C. Cowan, personal communication, May 21, 2010). Cowan does note that some of the organization's ranching members did splinter off, stating that members held pretty firm, but "we've had some members that joined OGAP" (C. Cowan, personal communication, May 21, 2010). Ms. Cowan describes these members as being "terribly upset" or being "unhappy with a lot of our policies" (C. Cowan, personal communication, May 21, 2010). The parent organization of New Mexico ranchers, to a lesser degree than their Wyoming counterpart, joined forces with environmental organizations, including the aforementioned OGAP. Cowan adds that the alliance was initially uncomfortable for New Mexico ranchers due to the entrenched "us versus them" lens through which ranching organizations have traditionally viewed environmental organizations (C. Cowan, personal communication, May 21, 2010). Cowan adds "It's formed interesting alliances that made me real uncomfortable to begin with. But then, you know, you learn that these people (OGAP) really don't have horns. Some of that stuff. So it was a growing experience for me in that direction" (C. Cowan, personal communication, May 21, 2010).

The Wyoming Stock Growers Association (WSGA) did not initially support reform efforts. Prior to the introduction of Wyoming's reform legislation, the WSGA "sat down and developed a plan...in partnership with the mineral industry" (J. Magagna, personal communication, March 23, 2009). According to Jim Magagna, "we put together a task force and developed a set of split-estate protocols, and there was a set of guidelines, both for the landowners and for CBM developers to say 'here are some steps you can take,' didn't provide the answers, but to enhance the communication, to enhance the understanding, to help them through the process. We provided mediation services as part of that to try and help these people work out some of these things [problems associated split-estate energy development] as much as possible on the front end" (J. Magagna, personal communication, March 23, 2009). In its effort to provide guidance to its members, the WGSA attempted to emphasize the building of relationships between CBM operators and ranchers prior to energy activities taking place (J. Magagna, personal communication, March 23, 2009). Efforts such as these, however, failed to appease some WGSA members' anger and frustration with what was occurring away from the task force. As John Vincent contends:

> The reason I think that you see these two groups [WSGA and energy] still working together is the oil and gas industry complains that they can't drill wherever they want to, whenever they want to, and the ag industry says we can't run sheep and cattle wherever we want to, whenever we want to. What happens, though, and where the disconnect happens is when you get down here actually on the ground. If you're not in the little club, so to speak, then you really are on the outside looking in, to the extent that a landowner tries to assert himself or herself, then you become a problem to the BLM and, you know, the other side of the equation. (J. Vincent, personal communication, March 16, 2009)

Vincent argues that the disconnect between members who were asserting themselves in their call for legislative intervention were left frustrated by what they viewed as the rhetorical appeasement of the energy industry by their parent ranching organization (J. Vincent, personal communication, March 16, 2009). As a result of their frustration, these ranchers splintered off and began to align themselves with environmental organizations.

The entrenched positions of the ranching-alliance are difficult to dislodge. As more ranching alignments were created, and more legislators began receiving calls from their constituents complaining about split-estate energy development, even Wyoming's traditional ranching organization, the WSGA, began working with environmental organizations. Ranching organizations' choice of which environmental organizations to work with remained limited to organizations ranchers did not perceive as a threat to their interests. Generally speaking, environmental organizations considered to have a conservation-oriented mission were those favored by the ranching industry. Environmental organizations' viewed as being preservation-oriented were still considered the enemy by ranching organizations in Wyoming and New Mexico (J. Magagna, personal communication, March 23, 2009; C. Cowan, personal communication, May 21, 2010). Jim Magagna and Caren Cowan both argue that because environmental organizations run the "spectrum," ranching must be selective in their alliance with environmentalists (J. Magagna, personal communication, March 23, 2009; C. Cowan, personal communication, May 21, 2010). Magagna comments that:

> I always want to distinguish the term 'environmentalist' from what I view as true conservationists, those who want to see the resource properly managed and cared for, as opposed to those who don't really want mineral development out there, [and] don't really want, at least public land, grazing, don't really want commercial recreation, the list could go on, but those are the three big ones. Whether it's formal groups or just individual citizens who think that locking up the land, or this notion of restoring to pre-European settlement conditions, which you hear about periodically, any of that, in a bigger scheme of things, is a far greater threat to our industry than what's happening with the mineral people. (J. Magagna, personal communication, March 23, 2009)

Caren Cowan reinforces this position by adding that in New Mexico, her organization, the NMCGA, takes the position that "OGAP is not the same as the Sierra Club. We don't view OGAP as somebody that's trying to get rid of grazing, where the Sierra Club is. So even we make those kinds of distinctions, but we have learned that it's our job to sit down with those people, whether we like it or not. You cannot expect us as an organization to go have them suing to get us off of the land on one hand and go hold hands with them on the other hand. It just doesn't work that way" (C. Cowan, personal communication, May 21, 2010).

The support of environmental organizations in ranching's efforts to implement legislative reform was difficult for ranching organizations to comprehend. As Caren Cowan notes, "Sometimes you kind of have to take a deep breath and scratch your head and say what did I miss in that? That this set of players that we're all on the same side" (C. Cowan, personal communication, May 21, 2010)? Jim Magagna argues that the working relationship with environmental organizations has, over time, proven beneficial to ranchers. Magagna comments that, "It's kind of interesting because yeah, as the shift started to take place, we were the beneficiaries of the fact that for awhile we were the target of the environmental community and then, suddenly, mineral development became the target, and we, in recent years, found ourselves to be the party that's being courted by both sides" (J. Magagna, personal communication, March 23, 2009). Mr. Magagna's suggestion that ranching is now courted by both environmental and mineral organizations was not the case, however, at the height of the legislative competition between ranching and energy.[3]

Assistance given to ranchers and their respective organizations by environmental organizations was done quietly. This is because, as Laurie Goodman argues, among members of LAW, "there is a truth of environmentalism as it relates to property and to our cause. But if casting our effort in environmental terms were to occur it would have killed us" (L. Goodman, personal communication, March 23, 2009). Casting the effort to reform split-estate energy development in environmental terms is different than casting those efforts in private property protection terms. This is because the protection of private property rights resonates with ranchers, while the environmental protection values of preservation-oriented environmental organizations remain unacceptable to ranchers. This means that the ranching-environmental alliance is subjective to the land management policy issue at hand. Ranchers have retained their traditional wariness of environmental organizations, even among those environmental organizations whose missions are considered by ranchers as relatively moderate.

From the perspective of the energy industry, the support of environmental organizations given to ranchers is unacceptable. As such, energy representatives used the newly formed ranching-environmental alliance in their opposition of ranching's efforts to reform split-estate energy development. The lobbying and negotiation activities of the ranching and energy industry were competitive. As tensions rose, a mutual state of mistrust took hold and communication broke down. Simply put, the conflict heightened to a point where ranching and energy organizations were no longer communicating with each other. Thus, discussions between representatives of ranching and energy development organizations that occurred during this period illustrate the degree to which the conflict between ranching and energy had heightened. For example, Caren Cowan tells the story of how the energy industry's representative used the ranching-environmental alliance as a weapon in his industry's opposition to NMCGA's support of New Mexico's Surface Owner Protection Act. Ms. Cowan relates that, at the request of the Governor of New Mexico, Bill Richardson, the two opposing sides were asked to close themselves off in a meeting room in the New Mexico state capitol building and begin communicating with each other. At the appointed time of the meeting:

> [The energy representative] leaned over the desk and got in my face and said, 'Just wait till your members find out you have brought in an out-of-state environmental group to carry your water.' I mean, he totally ignored the fact that [environmental representative] was even sitting there, and I blew up. I mean, I don't lose my temper very often. I lost my temper really bad. It's those kinds of things. I mean, it was totally unnecessary. I mean, you just, you don't treat other human beings that way. You may be mad that you're having to sit across the table from somebody, but you know, he knew how to push my button, and he did really well. I stormed out of the office and slammed doors that you

---

[3] Note: Surface Owner Protection Acts were debated from 2005 to 2007. As of 2010, Utah remains the only Western state without a Surface Owner Protection Act. The latest defeat for Utah's bill was in the 2010 legislative session.

could hear three floors down. (C. Cowan, personal communication, May 21, 2009; B. Gallagher, personal communication, May 21, 2009)[4]

During the course of debating State Surface Owner Protection Acts, representatives of the energy industry sought to use the alliance of environmental groups against ranching organization seeking to implement legislative reform. In the opinion of ranching organizations, the energy industry's opposition of their reform efforts left them no choice but to seek the assistance of the more conservation-oriented environmental organizations. The controversial strategy of seeking out the assistance of environmental organizations remained covert, known only to ranchers who had organized themselves outside their parent organizations.

Surface Owner Protection Acts: Ranching's Perspective

By 2003–2004, energy activities had displaced ranching activities as the predominate use of land and resources in states of the Rocky Mountain West. During the course of energy development's expansion, ranchers were hindered in their efforts to adequately respond to energy's ability to access their surface lands. Ranchers could not adequately address the problems associated with split-estate energy development because the right of developing the federally owned mineral estate is protected. Ranchers were frustrated by the BLM's response to requests for assistance in resolving problems associated with split-estate energy development and with the energy industry's lack of accountability as split-estate energy activities began impacting their ability to conduct ranching activities. Unable to resolve their problems amicably with the BLM or the energy industry, ranchers focused their efforts on enacting Surface Owner Protection Acts in order to protect their interests.

Ranching and energy organizations competed to protect their respective interests as Surface Owner Protection Acts were debated in Western states legislatures. On the one hand, energy officials believed the types of reforms being sought by ranchers were unnecessary. Ranchers, on the other hand, believed that the types of reforms they sought would restore equity and balance to their competing interest in land-use. As John Vincent notes, the competition between the ranching and energy industry over these reforms "was out and out war" (J. Vincent, personal communication, March 16, 2009).

Ranchers sought four types of reform in order to restore their equal footing with energy developers. First, ranchers requested that state legislators extend the notification time frame in excess of the federal standard. Second, they asked that guidelines be imposed on the types of accommodation and compensatory damages required by Surface Owner Agreements in excess of those required by federal law. Third, they requested that the state increase required bonding fees in excess of fed-

---

[4] Note: Unprompted, Gallagher, former Director of the New Mexico Oil and Gas Association (NMOGA), retold the same story and confirmed that it was he who had invoked Cowan's wrath.

eral standards. And finally, they asked for the requirement that energy development companies offer fair compensation for any loss of ranchers' potential property value (Interviews collectively).

Ranchers' willingness to confront energy developers is not the norm. A confrontation with the energy industry, in the opinion of most ranching representatives, is not in the best interest of the ranching industry. For ranchers who were determined to confront the energy industry, this meant having to separate themselves from their fellow ranchers. According to Jim Magagna, it was not that most ranchers lacked the willingness to engage in a fight; it was that "they are not able to" engage in a fight (J. Magagna, personal communication, March 23, 2009). Magagna argues that, ranchers are unable to engage in expensive confrontations with a more fiscally resourceful industry like that of energy. Magagna also notes that "it's not in their [ranchers] nature to want to spend time in court or the halls of the legislature" (J. Magagna, personal communication, March 23, 2009). Ranchers would, according to Magagna and other ranching representatives, rather be running their stock or engaging in other ranch activities (Interviews collectively). Magagna believes that while decision-makers "are going to be a little inclined toward the rancher as the little guy unless it is some big, huge, powerful person [energy]" (J. Magagna, personal communication, March 23, 2009). Magagna uses the analogy of a court hearing to illustrate ranchers' David-like position in the fight with the Goliath-type position of the energy industry when he suggests, "If the mineral company was able to bring in three high-powered lawyers to present a good case, and the rancher had to hire the neighbor, who's only been practicing for two years and isn't involved in oil and gas litigation and says, 'well, I can only afford to give you $2,000 to do the best you can for me,' it's pretty hard for a sympathetic judge to necessarily favor the rancher" (J. Magagna, personal communication, March 23, 2009). The same disparity holds true for ranching's ability to challenge the energy industry in the halls of state legislatures.

Today the ranching industry does not have the ability to strongly influence state-elected officials. This is because, as Caren Cowan contends, "We [the ranching industry] had a lot more boots in the legislature [in the past] than we have today. That's just the bottom line" (C. Cowan, personal communication, May 21, 2010). This does not mean that state legislators are unsympathetic to the plight of ranchers, but as Cowan notes, "Their [the legislators] ox wasn't being gored at this point, so they don't know how much blood was on the floor to get it [conflict] where it was" (C. Cowan, personal communication, May 21, 2010). Legislators and ranchers were hesitant to confront energy development because of the tremendous financial benefits that Western states derive from the production of energy. Jim Magagna comments that:

> Even if you compare Wyoming with some of our neighboring states, Wyoming derives tremendous value for our minerals. And I think, generally speaking, we have been friendly to mineral development. It is certainly not something we have been opposed to because we as agriculturalists benefit from that in numerous ways. Not only in terms of government services, but in terms of low taxes and other benefits as well. (J. Magagna, personal communication, March 23, 2009)

Ranching's ability to derive benefits from the development of energy notwithstanding, ranchers believe that federal laws favoring energy development create disparity between their ability to derive personal economic benefit from their surface activities. In the opinion of ranching representatives, that disparity is the direct result of the federal government's protection of energy development activities as an inherent property right. Ranchers oppose the disparity of energy leases having the rights of property, while grazing leases are not recognized as having the rights of property. Caren Cowan argues that ranchers understand the property right of an energy lease assures that the energy resource will be developed and brought to market, but that "as we were fighting to get this bill [Surface Owner Protection Act] passed there was a lot of conversation that went on that it [mineral estate] was the dominant estate, and that we couldn't do anything to change that because it would harm the federal law. In other words, we could not flatly deny access" (C. Cowan, personal communication, May 21, 2010). Ranchers could, however, attempt to enact measures that would extend or exceed federal guidelines regarding mandates of providing surface owners with a notification of access, accommodation and compensation for surface disturbance, and bonding fees.

Federal regulations require that 45-day notice be given to landowners of a company's intent to access the surface estate and conduct energy activities. It is common practice to provide notices via certified mail. Once delivered, the 45-day clock begins to run. If the landowner does not respond to the notice, the permission to access and begin energy development activities is implied. Caren Cowan confirms that these types of notifications were often put aside by ranchers with the understanding that the land-man with whom the rancher had always dealt with would eventually show up at the door and work a deal out for access and accommodation (C. Cowan, personal communication, May 21, 2010). Cowan verifies that it was often the case that the land-man would not show up and that that the 45-day time limit for a response would pass (C. Cowan, personal communication, May 21, 2010). The result was, according to Cowan, that instead of the land-man coming to the ranch, the rancher would, eventually, be met by trucks coming through the gate and across the surface of the ranch (C. Cowan, personal communication, May 21, 2010). Jim Magagna adds that energy development's rapid expansion led "to more shortcuts being taken by the mineral companies in their effort to get out there and get a lot done in a hurry" (J. Magagna, personal communication, March 23, 2009).

The increasingly regular occurrence of these types of instances motivated ranchers to request that the state enact statutes that would extend the time frame in which ranchers could respond to the access and development notification. Jim Magagna describes the issue of adequate notification of access and the intent to develop as being one issue among many in "some of this head-to-head [argument between ranchers and energy developers] on specific issues, on some of the issues that were eventually addressed in split-estate" (J. Magagna, personal communication, March 23, 2009). In this sense, ranchers requested that, in addition to extending the time frame of a notification, states require energy companies to notify split-estate property owners of their intent to purchase an energy lease if the lease was located within the boundaries of a split-estate property holding. The purpose of asking for advanced notification of an energy developer's intent to purchase a lease was, according to Ms. Cowan, to give the rancher a "time frame to bid against them" (C. Cowan, personal communication, May 21, 2010). The response of the energy industry to these proposals was that "they got really angry" (C. Cowan, personal communication, May 21, 2010). Cowan remarks that the energy industry's stance was that additional notice requirement "doesn't do anything" to alleviate problems associated with inadequate notification (C. Cowan, personal communication, May 21, 2010). Cowan relates that when her organization inquired "if it doesn't do anything to you, then why enter into this fight" (C. Cowan, personal communication, May 21, 2010)? Cowan notes during the course of a meeting with a representative of energy the response to her question was that, "energy got really mad and turned around and walked off" (C. Cowan, personal communication, May 21, 2010).

Federal law only stipulates that split-estate landowners be compensated for the loss of crops—including plants and grasses associated with grazing—and existing structures. As noted earlier, if accommodation and compensation for losses resulting from energy activities disturbance of the surface are not addressed in the Surface Owner Agreement, energy companies are not required by law to compensate the landowner after the damage has occurred. This type of issues was most problematic to ranchers. As noted earlier, split-estate property owners were often unfamiliar with the types of accommodation and compensation issues they should be addressing with the energy company's representative. Hindered by their lack of knowledge and the inadequacy of the information that was made available, split-estate property owners' sustained damages of one type or another that resulted from energy development activities not addressed in Surface Owner Agreements. As a result, the most common problems associated with the development of split-estate energy development were those associated with unaddressed issues of accommodation and compensation.

Ranchers employed three justifications for their request that state legislatures enhance energy developers' monetary accountability to ranchers for inconvenience and losses. The first justification ranchers made concerned their request that the state impose specific guidelines regarding the types of accommodation and compensation issues that energy developers were required to address in Surface Owner Agreement contracts. As John Vincent argues:

> The type of conflict that typically arose wasn't so much that the farmer or rancher wanted to stop the drilling; they weren't opposed to the idea that the oil company was there improving their estate or developing it or any of those types of things. Where the rub always occurred was pretty much always in the drilling location. The disputes that typically occurred were: Why are you [company] putting it in the middle of the field rather than on the edge of the field? Why aren't you hauling out all of those cuttings from that reserve pit and getting them out of here? Why aren't you locating your production facilities off to the side of the field? Those kinds of issues. (J. Vincent, personal communication, March 16, 2009)

The ranching industry's justification for states enacting a requirement that energy companies address issues such as these during the course of negotiating Surface Owner Agreements was intended to offset the lack of specificity in federal laws and regulations regarding issues accommodation and compensation. Vincent adds that "you can try to determine damages, but then the rub became that the damages that were permitted under the law were so miniscule that basically it was a taking without compensation. Outbuildings and crops, and irrigation improvements" (J. Vincent, personal communication, March 16, 2009). Vincent contends that the energy industry's response to ranching's request of clarifying the types of accommodations and compensation addressed in their negotiations with split-estate property owners was met with stern opposition (J. Vincent, personal communication, March 16, 2009). Vincent comments that "the oil companies fought hammer and tong about the notion that you [rancher] could have a determination of or recover damages in the difference in the value of the land before and after entry. Of course, the oil companies are trying to limit that determination to just that acre, or whatever they say is involved by their activity" (J. Vincent, personal communication, March 16, 2009).

Caren Cowan suggests that uniformity of the types of accommodations made and the compensation amounts considered in a Surface Owner Agreement should act as a "guideline" for the negotiation process (C. Cowan, personal communication, May 21, 2010). She argues that because "there's always what ifs that nobody can anticipate," uniformity should guide the course of the negotiation process, and that "having some guidelines would be very helpful for people" (C. Cowan, personal communication, May 21, 2010). Cowan clarifies that "if it's a guideline, I don't think that it would be good to have something that there's no deviation from. I'll get calls [from split-estate landowners asking] what do I need to do? Do you have a kind of template? Where do I start" (C. Cowan, personal communication, May 21, 2010)? Jim Magagna concludes that "really taking the time to sit down and negotiate, some of the things that in an earlier day the mineral operator and the landowner worked out over the kitchen table. There wasn't that kitchen table atmosphere anymore; it was here's an agreement, sign this. The drilling rig will be here day after tomorrow-type approach" (J. Magagna, personal communication, March 23, 2009).

The second justification ranchers made concerned their request that the state impose higher bonding fees for energy development activities. Federal law limits the amount of required bonding fees for energy developers to access and develop energy resources. As previously discussed, federal bonds range from site bonds of $1000, to blanket bonds of $150,000 for nationwide energy development projects.[5] Ranching organizations consider these bond fees inadequate to recover potential economic losses they may suffer or the costs of reclamation they may incur once energy activities conclude. Ranchers are particularly concerned about the inadequacy of federal bonding requirements because of their legal incapacity to prohibit energy companies from accessing and developing the energy leases located on split-estate properties.

Ranching's justification for the state's imposing higher bonding fee requirements from energy companies seeking to engage in energy development activities was intended to offset the low dollar bond amounts established by the federal government. Caren Cowan suggests that ranchers feel as they can neither prohibit energy development activities from occurring nor can they negotiate the terms of

---

[5] Note: See generally discussion in Chap. 4

a bond (C. Cowan, personal communication, May 21, 2010). Cowan argues that, "they [energy companies] were just [posting bond with the BLM] and coming on" (C. Cowan, personal communication, May 21, 2010). In turn, according to Cowan, "They [ranchers] feel like they're negotiating with one hand tied behind their back" (C. Cowan, personal communication, May 21, 2010). Ranchers, Cowan notes, believed that "the bonding was too low and it [lack to negotiate the price of a bond] handicapped them in that way" (C. Cowan, personal communication, May 21, 2010). Cowan suggests that ranchers would "prefer that the bond be $200,000 as opposed to $25,000 because outside of the BLM there is nobody to hold them [the energy industry] back" (C. Cowan, personal communication, May 21, 2010). Jim Magagna adds that "It's the federal mineral, and certainly they [energy developer] can bond on. But beyond that, I guess arguably from the federal perspective, at one time, the federal government had it all, and through the Stock Raising Homestead Act they issued me [rancher] a patent for the surface and so if they issued it, it's subject to terms and conditions under which they issued it, or at least now claim that they issued it, which [in terms of the severed mineral estate] is an absolute right to develop the minerals" (J. Magagna, personal communication, March 23, 2009). Magagna concludes that given the limits of federal law and regulation:

> One of the things we really preach to ranchers is if you're in an area where CBM development is likely or inevitable, develop a plan looking 5, 10, 20 years down the road for your ranch; what you want. And then when the land-man comes knocking on your door, instead of he has a plan; he knows his plan of development for the well, for the coalbed methane, and you don't know what your plan is, so you're at his mercy. But if you've got a plan to put on the table and say, 'well, hers what my ranch plan is for the next 20 years, and how can you, through your CBM development, help me achieve that, or at a minimum, not hinder my reaching that goal, that you and I can do some positive things together. (J. Magagna, personal communication, March 23, 2009)

The third justification ranchers made concerned their request that the state require energy development companies to offer fair compensation for any loss of ranchers' potential property value. Ranchers justified their request by advocating that the state's legal requirement for the compensation for any loss of potential property value was intended to protect their ability to divide, or parcel, their property holdings in order to derive economic benefit from the land's value as real estate (Interviews collectively).[6] The energy industry, in the words of John Vincent, fought this proposal "hammer and tong" (J. Vincent, personal communication, March 16, 2009). Representatives of ranching agree that their proposal to hold the energy industry accountable for the loss of the ranchland's potential value as a real-estate holding was a "nonstarter" (L. Goodman, personal communication, March 23, 2009). Representatives of the energy industry viewed ranching's request for the potential of economic loss of value in the land as being "speculative," and they opposed the notion each time the topic of real estate value was brought to the negotiating table (L. Goodman, personal communication, March 23, 2009).

---

[6] Note: See generally discussion in Chap. 4.

When the dust settled, the states of New Mexico, Colorado, and Wyoming did enact Surface Owner Protection Acts.[7] Jim Magagna concludes that at the end of the legislative competition, ranchers provided legislators the opportunity to create a "better balance" between ranching and energy developers. Magagna adds that, "there was a place for policymakers to step in and create a little better balance and avoid these individual negotiations so frequently ending up in courts, or if they do end up in the courts, giving the courts a little something in order to allow them to provide some balance" (J. Magagna, personal communication, March 23, 2009).

The conflict and competition that occurred between ranching and energy established unprecedented wariness between the two interests. The ranching-energy alliance has been disrupted as the two industries went "eyeball-to-eyeball" with one another as each interest sought to defend their industry's interests (Interviews collectively). The legislative battles ranching and energy's organizational representatives had engaged in were fights to control the future destinies of their respective members. On the one hand, the energy industry—relying on the legal dominance of the federally owned mineral estate—sought to defend the status quo of split-estate energy development. In doing so, they sought to control ranching's ability to reform split-estate energy development. Ranchers, on the other hand, had sought to protect themselves from what they viewed as a fundamental unfairness that was greatly impacting their ranches and their communities.

Annexing the BLM's Land-Use Subgovernment: Ranching's Perspective

Representatives of ranching are uniform in their belief that the BLM's land-use subgovernment has been superseded by the energy industry. Their opinions demonstrate ranching's deference to energy development interests within the hierarchical structure of the subgovernment's network of interests. As ranching's representatives all note, they believed that their industry's displacement within the BLM's land-use subgovernment was only a matter of time. Representatives of ranching organizations are not shy in expressing the opinion that the federal government, and in particular the BLM, and environmental organizations are not friendly to their interests. They remain collectively dismayed at the energy industry's assault on their attempt to protect their interests from harm. But that does not stop them from agreeing that the interests of ranching remain intertwined with those of the energy industry. Ranching representatives concede that times have changed for their industry and that a number of variables have, over the course of time, played a role in their fall as the predominant voice within the BLM's land-use subgovernment.

---

[7] Note: Wyoming Surface Owner Accommodation Act (2005); New Mexico Surface Owner Protection Act (2007); Colorado Surface Owner Protection Act (2007)

Energy derives greater political, economic, and social benefits than does ranching. Ranching's deference to energy development occurs because as Caren Cowan argues, "It's been a long, long time since grazing was the most economically valuable piece of that [subgovernment] because nobody puts a value on the stewardship and the care of the land" (C. Cowan, personal communication, May 21, 2010). Cowan's argument is premised on the belief that "the BLM is tasked with making as much money as they can" (C. Cowan, personal communication, May 21, 2010). Ranching representatives acknowledge the important role domestic energy development plays in the political, economic, and social well-being of the United States. But they also view the legal and political protection of the energy industry is at their expense. Among the ranching organizations of the Rocky Mountain West, the view is that the legal and political defense of energy development is unbalanced. Cowan concludes that the BLM takes its marching orders from its elected political masters: "We think at this juncture that the BLM does pay a lot more attention to the oil and gas industry, and that's said with all due respect to the good people I work with in the BLM every day. I think the mandate from D.C. down is that way" (C. Cowan, personal communication, May 21, 2010).

Western states' revenues are dependent on energy development. The economic return of energy development far exceeds the return ranching revenues to state budgets. The economic benefits created by the expansion of domestic energy development are important to governmental decision-makers. It can, however, lead to the development of cozy relationships that are not beneficial to ranching's or the public's interests when conflict occurs. As John Vincent argues, "I just think that when you have an industry that grosses 34 Billion dollars, self-reports 34 Billion dollars in revenue in a state like Wyoming where there are five or six hundred thousand people it's kind of like ignoring the elephant standing in the living room. That industry is going to have an influence on local government, county government, state government, the BLM" (J. Vincent, personal communication, March 16, 2009). Vincent notes that the energy industry's influence on people, like that of government entities, can be "overwhelming" (J. Vincent, personal communication, March 16, 2009). Vincent comments that, "they [energy industry] can hire rafts and rafts of lawyers and lobbyists, whatever, that are just up in people's faces all of the time. They're in their stroking them, they're in there, you know, let me help you help us get that permit. What the oil and gas industry does is curry favor with the people they think will be their allies" (J. Vincent, personal communication, March 16, 2009). Vincent concludes that when those persons or groups, such as disgruntled ranchers, begin to complain to their elected officials, the result is "the legislature sitting there thinking, 'what do I do now'" (J. Vincent, personal communication, March 16, 2009)?

Energy's annexation of the BLM's land-use subgovernment away from ranching began as domestic energy development increased. Ranching's loss of land-use decision-making control within the BLM's subgovernment coincided with the impact political willpower, energy markets, and technological advancements on the energy industry's capacity to expand their operations across the Rocky Mountain West. Caren Cowan argues that, "I look back to [Secretary of the Interior]

Babbitt's[8] Range Land Reform and some of the stuff that he did. When they [federal government] did away with grazing advisory boards basically, we lost a voice that we had through all of this [conflict with energy] and I tend to believe that had we still had those grazing advisory boards that we might have had more voice in what went on" (C. Cowan, personal communication, May 21, 2010). Cowan considered the BLM's grazing board replacement with Regional Advisory Councils (RAC) as being effectively "ceremonial" (C. Cowan, personal communication, May 21, 2010).[9] But, she is also quick to point out that the ceremonially nature of the RAC decision-making process does not deter ranching from taking its seat at the table because "at least there's a process, at least there's a voice if you really have to do something. If you don't go, they say, 'you didn't want to play. You had an opportunity to play and you chose not to'" (C. Cowan, personal communication, May 21, 2010). Cowan also contends that ranching's lack of resources hinders its ability to more fully engage in BLM Resource Management Planning (RMP) (C. Cowan, personal communication, May 21, 2010).

Ranching at one time dominated the network of interests within the BLM's land-use subgovernment. Ms. Cowan regards ranching dominance as a thing of the past when she comments "You know, back in the '30s or '40s maybe, in the very beginning perhaps, but in my lifetime? No" (C. Cowan, personal communication, May 21, 2010). Jim Magagna responds that ranching's dominance was diminished, if not lost, with the enactment of the Federal Land Policy and Management Act of 1976 (FLPMA) (J. Magagna, personal communication, March 23, 2009).[10] Magagna hastens to add that viewing ranching through the lens of a pre-FLPMA iron-triangle, "Grazing was very powerful at one point in time. I'm not sure that back at that time I ever thought of grazing and minerals as being together because minerals were simply less of a factor" (J. Magagna, personal communication, March 23, 2009). Magagna comments that, "Prior to the creation of the BLM, grazing on these lands existed. Most of the policy centered around grazing because that's basically all there was out here on most of the BLM lands. There wasn't anything else. So certainly in that sense there was a dominance" (J. Magagna, personal communication, March 23, 2009). Magagna adds that in the face of energy development's expansion across the West "livestock grazing has clearly been allowed to continue, but pretty much at a low level of attention from the Bureau [BLM] itself, a low level of importance. I mean few people outside the ranching community know what you're talking about when you talk about the Taylor Grazing Act today" (J. Magagna, personal communication, March 23, 2009). As a result of the employment of FLPMA and the mandate of multiple-use, ranching's dominant position within the interest network of the BLM's land-use subgovernment eroded.

---

[8] Note: Cowan is commenting on Secretary of the Interior Bruce Babbitt under former President William J. Clinton

[9] Note: Under the Federal Land Policy and Management Act of 1976 BLM Regional Advisory Councils (RAC) were created to replace the grazing commissions of the General Land Office and U.S. Grazing Service

[10] Note: See generally discussion in Chaps. 1, 2, and 3

The BLM does not direct the same amount of resources toward ranching as it does energy development. It is the shared opinion among representatives of the ranching industry that a shift in the subgovernment led to a shift in BLM policy and resources. Representatives of the ranching industry comment that "the mineral industry has the disproportionate amount of attention of the BLM" (Interviews collectively; J. Magagna, personal communication, March 23, 2009). According to ranching representatives, it's not just the attention of the BLM favoring the energy resources, but more of the BLM's administrative resources are directed at energy development as well (Interviews collectively; J. Magagna, personal communication, March 23, 2009). The ranching industry's loss of prominence within the BLM is attributable in part to what Jim Magagna refers to as "a totally different historic setting that doesn't repeat itself today. Because when we had so-called control of it, it wasn't that we were demanding their [BLM] attention, their resources. It was, some would argue, it was sort of a free pass" (Interviews collectively; J. Magagna, personal communication, March 23, 2009).

Conclusion: Annexation of a Subgovernment

Ranchers never wished to halt the development of domestic energy resources (Interviews collectively). This is a common theme repeated among representatives of ranching's interests. However, as domestic energy development spread across the Rocky Mountain West, more and more ranchers were affected. The negative effect of domestic energy's expansion was a particularly onerous on ranchers who own split-estate properties. With increasing numbers of split-estate energy properties being developed, the short-term impact to ranchers was the disparity in their ability to fairly negotiate terms of access and development with energy developers. The long-term impact to ranchers was their inability to recoup economic losses from increased development activities occurring on their surface lands. Because impacts such as these were not being addressed by the BLM or the energy industry, problems associated with split-estate energy development triggered conflict between ranchers and energy development interests.

With each new report of a problem occurring, tensions between the ranching and energy industries heightened. In turn, because ranchers' problems with split-estate energy development centered on the inequities that had been created by federal laws and regulations guiding split-estate energy development, ranchers and their traditional lobbying organizations sought the protection of their state legislatures.

Ranchers seeking protection of state law advocated for the reform of split-estate energy development by petitioning their state governments to enact Surface Owner Protection Acts. In the legislative battles that ensued, each industry sought to control and protect its industry's predominant use of the land and resources. Ranching and energy development organizations lobbied their states' elected officials in a manner that would prove most valuable to their members' interests.

At the conclusion of these legislative battles, it is clear that the alliance that had existed between these two powerful and resource-rich stakeholders in the BLM's land-use subgovernment had been disrupted. Policy control of the BLM's public land decision-making subgovernment has shifted. Policy over the manner in which governmental decisions are crafted regarding the use of federally managed public lands is now controlled by the energy industry.

References

Cowan, C. (2010). Director of the New Mexico Cattle Growers Association (NMCGA). Interview conducted: May 21, 2010; Albuquerque, NM.

Goodman, L. (2009). Chief Legislative Lobbyist for the Landowners' Association of Wyoming (LAW). Interview conducted: March 23, 2009; Cheyenne, WY.

Magagna, J. (2009). Director of the Wyoming Stock Growers Association (WSGA). Interview conducted: March 23, 2009; Cheyenne, WY.

Vincent, J. (2009). Legal Counsel to the Landowners Association of Wyoming (LAW), former mayor of Riverton, WY. Interview conducted: March 16, 2009; Riverton, WY.

Chapter 8
Conclusion

Abstract Legal protection of the mineral estate is disproportionate to the legal protection of the surface estate. Enforcement of the mineral estate's dominance over the surface estate is furthered as federal regulations guide split-estate energy development. This body of federal law and regulation guide the BLM's management and oversight of split-estate energy development. It establishes the federal government's prevailing interest in developing the federally owned mineral estate. This body of law and regulation also mandates the BLM protect the government's interest in developing the mineral estate. It is a mandate that reflects the intent of government to serve the public welfare by protecting its ability to provide energy resources to the nation. Findings indicate that the government's legal and regulatory protection and development interests in split-estate energy resources contradict fundamental principles of property ownership and environmental stewardship. The antiquated nature of federal law and regulation controlling the development of split-estate energy resources are at odds with the legal, political, economic, and technological realities of modern-day domestic energy development. The shared understanding expressed by government, energy, and ranching officials supports this conclusion. Development of a non-traditional energy resource would not have been possible without a "perfect storm" of legal, political, economic, and technological conditions all coming together within a relatively short period of time. These events established the conditions for political upheaval in the BLM's land-use subgovernment.

Keywords Bureau of Land Management · Department of Interior · Bureaucracy · Split-estate energy development · Federal law · Federal regulation · Executive power · Subgovernment · Natural Resource Policy

The Energy Industry Dominates the BLM

Fluctuations in three principle factors led to the expansion of domestic energy development in the Rocky Mountain West. First, shortages in energy resources created greater demand for energy supplies, which led to a rise in the price of

© Springer Nature Switzerland AG 2019
R. E. Forbis Jr., *Altered Policy Landscapes*,
https://doi.org/10.1007/978-3-030-04774-0_8

energy supplies. Second, technological advances furthered the ability of energy companies to develop hard-to-access energy supplies, including non-traditional energy resources such as coalbed methane (CBM). As energy prices increased, and the use of better technology became widespread, development of energy resources was again profitable for energy companies. Energy companies sought to profit from traditional and non-traditional energy resources and creating a desire among energy developers to expand domestic energy development. Third, elected officials, responding to the need for expanding domestic energy development, took steps to improve the ability of energy companies to expand their energy resource development activities. Therefore, fluctuations of energy markets, energy technology, and political willpower converged to facilitate rapid expansion of domestic energy development.

The Bush Administration succeeded in its use of executive power in strategic pursuit of its expanding domestic energy development. They were able to shift the energy policy of the Bureau of Land Management (BLM). This is because the BLM, in response to the president's political directives, shifted its energy policy by implementing changes to the administrative procedures that guide the development of domestic energy resources. Congress responded to this shift by increasing the administrative resources of the BLM in support of the president's policy objective of expanding domestic energy development. Therefore, changes in the executive branch led to changes in domestic energy policy.

Domestic energy development activities increased in states of the Rocky Mountain West. Three to five percent (3–5%) of all domestic energy development from 2001 to 2009 took place on split-estate property in the states of New Mexico, Colorado, and Wyoming. A split-estate is defined by federal law as a parcel of private property where rights of the privately owned surface estate are severed from the rights of the federally owned mineral estate. As energy development activities on split estates expanded, problems associated with those activities negatively affected split-estate property owners, ranchers, and homeowners. In turn, as energy development-related problems between split-estate landowners and energy developers multiplied, landowners—particularly ranchers—sought to protect their interests by lobbying state lawmakers for enactment of Surface Owner Protection Acts. During deliberation of Surface Owner Protection Acts by state lawmakers, organizations representing the interests of ranchers and energy developers competed for control of the federal land management policy environment. Therefore, changes in domestic energy policy triggered heightened conflict and competition between formerly allied, strong, and resource-rich members in a public lands subgovernment.

Surface Owner Protection Acts were eventually enacted by the states of New Mexico, Colorado, and Wyoming. These acts were intended to heighten the accountability of energy developers in the development of split-estate energy resources. The legislation increased federal notification standards and bond fees. Ranching organizations supported these measures. Energy development organizations opposed these measures. The confrontation between ranching and energy development over the intent of the legislation and the means proposed to achieve the legislation's intent divided the formerly allied interests of the ranching and energy industries. Evidence suggests that as the confrontation between ranching and energy unfolded in state legislatures, conflict and competition between the two interests heightened. Evidence also suggests that as the conflict and competition heightened over the enactment Surface Owner Protection bills, the central concern of each side was to reform or protect federal law and regulation concerning split-estate energy development. Therefore, heightened conflict and competition between former subgovernment allies led to a shift in policy control of a public lands subgovernment.

Surface Owner Protection Acts did not reform federal laws establishing the mineral estate's legal dominance over the surface estate. While energy developers are held more accountable in some states, the BLM's administrative procedures guiding split-estate energy development are unchanged. Federal laws and regulatory procedures are unchanged because federal energy policy is largely unaffected by state Surface Owner Protection Acts. The inability of states to reverse or affect change to federal energy policy suggests that central administrative authority for split-estate energy resource development and its activities still rests with the BLM. Therefore, the BLM is no longer a rancher-dominated agency, but is now an energy-dominated agency.

Presidential Control over Subgovernments

Evidence indicates there are three principal coalitions of stakeholders interacting within the interest network of the BLM's land-use subgovernment: energy, ranching, and environmental. These interests comprise what might be referred to as the triumvirate powers of the BLM's land-use subgovernment. These three powerful and resource-rich interests have, over time, competed for control of the federal land management policy environment which resulted in the establishment of a hierarchical relationship within the interest network of the BLM's land-use subgovernment. Over time, this hierarchy of interests became entrenched. Ranching retained its influence among the two corresponding networks of actors (executive agencies and congressional committees) that compose the BLM's subgovernment. In doing this, ranching controlled land management budgets and policies. Simply put, the ranching industry managed to defend its domination of the BLM's subgovernment for decades, and in doing so, it retained control over the policymaking environment from the years of Western expansion to election of George W. Bush and Vice President Richard B. Cheney in 2000.[1]

---

[1] Note: This conclusion does not imply that the 2000 election of the Bush Administration was the only factor in the demise of ranching's power over the BLM's subgovernment. Other factors, such as the Taylor Grazing Act of 1934 and the Federal Land Policy and Management Act of 1976, along with generational movement of ranching families from rural ranchlands to metropolitan population centers are just a few among the many factors that have led to ranching's inability to be the dominant force in BLM land-use decision-making.

The analysis of archival historic documents combined with investigative interviews of elites presented here indicates that the 2000 election of the Bush Administration resulted in a change in the BLM's energy policy. This change disrupted the established subgovernment of the BLM's land management policy environment. The analysis also indicates that, in its current manifestation (2010), the established hierarchical order among the principal interests in the BLM's network of interest groups is energy first, ranching second, and environmental third. This disruption is indicative of the Bush Administration's pursuit of expanding domestic energy development. Therefore, based on the evidence presented here, we can conclude that unilateral actions taken by the president can have the impact of disrupting established subgovernments.

The finding that a president can disrupt established subgovernments is based on three conditions. First, a president must unilaterally exercise executive powers in strategic pursuit of a policy objective. Second, a president who exercises executive powers in this manner must have the support of Congress. Finally, a president must also have the support of congressional committees charged with oversight of the policymaking environment in which the policy objective is being sought.

Analysis of the development of energy policy during the Bush Administration provides compelling evidence that these three conditions promote and facilitate substantial shift in policy and subsequent change in a land management agency subgovernment. In the case of energy development policy, the first condition was met when President Bush exercised his executive powers unilaterally in strategic pursuit of expanding domestic energy development. The second condition was met when the Republican-controlled Congress supported the administration's policy objective of expanding domestic energy development. The third condition was met when congressional subcommittees overseeing the land management policies of the BLM supported the objective of expanding domestic energy development.

The implication is that with every change of administration, there can be a corresponding change in pursuit of a different policy objective. The success changing direction, however, is dependent on the three conditions described above. Should any of the three conditions not be met, an administration's ability to achieve its preferred policy objective is substantially diminished. All three conditions must be present in order to disrupt an entrenched policymaking subgovernment. Policy objectives not favored by a subgovernment's dominant interest group will meet with failure unless that interest group is displaced.

The implication for a public lands agency like the BLM is that change in the executive branch can affect a corresponding shift in land management policy. More generally, shifting an administrative agency's existing policy environment is dependent on the desire and willingness of a newly elected administration to strategically employ executive powers in order to initiate that change. In doing this, the president must focus efforts on the existing body of federal law and administrative procedures relative to the policymaking environment the administration targets for change. Finally, the policy environment targeted for change by the president, as well as the president's objective for effecting that change, must have the support of a like-minded Congress and the subcommittees that compose the congressional network of actors of the policy subgovernment.

These findings suggest that over time, those concerned with developing a better understanding of a president's ability to impact an established subgovernment's policymaking environment are best able to do so during periods of political change and upheaval. The objective of this line of inquiry would then answer the question of whether it was possible for a president to disrupt a subgovernment and if so, to what effect? Doing so presents an opportunity for researchers to better determine the degree of effect a president's unilateral use of executive powers in strategic pursuit of a political objective has on a policymaking environment. As noted here, by his actions, President Bush impacted a relatively stable land management agency subgovernment. The degree to which his action impacted the policymaking environment is measured by his administration's ability to dislodge the dominant interest within the subgovernment and replace it with another interest more suitable to achieving the president's political objective. President Bush's impact on the policymaking environment of the BLM could not have occurred without the support of Congress and congressional subcommittees.

Questions remain concerning the level of effect the supportive actions of Congress and congressional subcommittees had on the president's ability to impact the BLM's policymaking environment. This is a line of inquiry that needs further investigation. Conducting a research effort of this type would assist political science researchers to develop a better understanding of the impact the congressional network of actors has on subgovernments. Future research findings may also lead to a better understanding of the degree to which Congress and in particular congressional subcommittees have for effecting political control over a subgovernment. One possible path of inquiry might be the role the budgeting process plays in disrupting a relatively stable subgovernment policy environment in a period of political disruption, conflict, and competition.

Energy Developers-Ranchers-Environmentalists

The forged alliance of ranching energy is unlikely to completely unravel. Retention of commodity-oriented use of federal lands and resources is important to both groups. Additionally, both industries view organizations with environmental or recreation-oriented uses of federal lands and resources as interlopers. They view these outside interests with suspicion and hostility.

Many ranching and energy interests share the opinion that environmental organizations are meddlesome. This opinion is premised on the suspicion that if ranchers should become too allied with environmental organizations, environmental organizations would "divide and conquer" (Interviews collectively). The shared belief among ranching and energy interests is that the issue of split-estate energy development activities might be "the wedge issue" environmentalists have been longing for (Interviews collectively). Thus, the ranching-energy alliance's shared belief that environmental organizations are the enemy is relatively intact. However, among some ranching interests, the entrenched argument of "us against them" has begun to weaken. This finding suggests that should a stewardship issue like split-estate energy development emerge, ranching interests would be hard pressed not to forge stronger alliances with most environmental organizations. Therefore, a new alliance could take shape within the BLM's land-use subgovernment. This is of course dependent on the steady decline of ranching in the West and thus the gradual weakening of their influence generally.

Based on the research conducted for this book, the struggle for policy control of BLM's policymaking environment weakened the ranching-energy alliance. However, the interests of ranching-energy alliance remain intertwined because representatives of ranching and energy frame their relationship generally as being "mutually beneficial" (Interviews collectively). This positive orientation suggests that even during times of disruption, the forged alliances that compose the strong corners of subgovernments are never fully disentangled. Thus, it remains true that any permanent displacement of an existing hierarchical array of actors within a relatively stable subgovernment is difficult to achieve.

These findings lend support to the argument that during times of political upheaval, typologies of conflict are identifiable and definable (McCool, 1989, 1990, 1995, 1998). Furthermore, it lends support to the argument that during a period of political upheaval, a pattern of conflict emerges (McCool, 1989, 1990, 1995, 1998). Evidence and findings of this case study account for (1) the factors affecting the relative power of the BLM's subgovernment's participants, (2) the conditions and factors that provoked change in the BLM's subgovernment, (3) the variables that affected the level of integration between the BLM's subgovernment and its external environment, and (4) uncovered the democratic implications of the finding that the BLM's subgovernment policy environment is controlled by the energy industry.[2] The ability to account for these factors, conditions, variables, and implications lends validity to use of the subgovernment model as a lens of inquiry.

The validity of this conclusion is, however, limited because the analysis of subgovernment participants is limited. One limitation, mentioned previously, is the absence of state lawmakers from the study, who were affected by expansion of split-estate energy development in their respective states. Another limitation is the absence of homeowner and home development associations who were affected by the expansion of domestic energy more generally and split-estate energy specifically. Finally, a more significant limitation to these conclusions is the absence from this study of environmental organizations who participated in the conflict and competition over control of the BLM's policymaking environment. The roles and actions of these groups of actors represent further lines of inquiry that require more research in validating the findings of this research as well as the usefulness of the subgovernment model.

---

[2] Note: See generally McCool, 1989, 1990, 1995, 1998.

The Bureau of Land Management

Federal law formally restrains the discretionary decision-making power of BLM administrators. Historically, administrative discretion is restrained because development of the federal mineral estate is the preferred use of federal lands and resources. Contemporary government's preference in developing the mineral estate stems from the economic benefits government derives from energy resource development. Simply put, the economic benefits that federal and state governments derive from energy development outweigh the economic return from surface development activities such as grazing or recreation. The legal and economic disparity between mineral and surface estate development activities make balanced use of federal lands and resources difficult for BLM administrators to sustain. This difficulty suggests that the formal culture of the BLM is rule-bound. This condition undercuts the authority of BLM administrators to intervene on behalf of split-estate property owners. This finding suggests that by restraining the discretionary decision-making authority of BLM administrators, the legal dominance of the mineral estate undermines the public's expressed desire for multiple-use approach in the development of public lands and resources.

The informal culture of the BLM has shifted as well. This unexpected finding is supported by evidence that BLM personnel were no longer largely representative of Western cultural or agricultural backgrounds. This sentiment was articulated repeatedly by government, energy, and ranching actors. For example, a senior DOI appointee commented that, "When you go to BLM today, you don't find native westerners in many instances in these key BLM slots. For instance, the guy who oversaw the oil and gas development in Wyoming was from New York or Vermont" (Unnamed DOI political appointee, personal communication, May 26, 2009). John Vincent confirms the shift in BLM personnel and notes that the shift has implications for non-energy-related interests interacting with the BLM, noting:

> At least on a local level in terms of getting a drilling permit issued that really is done between two or three people: A permitting analyst from the company and a supervising engineer over in Lander. Those two people have to trust one another. They have to believe that the information that the company guys are providing is reliable, and … and so what happens is that the, the landowner doesn't have a place at the table. I mean they're not even there to say wait a minute, you need, did you think about this, or did you think about that? And the guy that you're talking to in the BLM is probably a petroleum engineer. (J. Vincent, personal communication, March 16, 2009)

This finding needs further investigation. Further research would assist in developing a better understanding of the impact non-Western natives have on the informal culture of BLM field offices. It would also help uncover how administrators interact with various interests who use federal lands and resources and the effect of those interchanges on citizen participation in the BLM decision-making process. Finally, a research effort of this type would be a helpful investigation of the corporatization of the BLM. One possible line of inquiry into a question of this type might be an exploration into the educational and professional backgrounds of BLM personnel or the types of services contracted out by the BLM.

Collectively, BLM administrators express that their land management policy decisions are directed and shaped by federal law. BLM administrators interviewed for this research were uniform in emphasizing that federal law dictates that the development of the mineral estate is the preferred use of lands and resources. Bound by those laws, BLM administrators are restrained from intervening when conflicts arise from split-estate energy development. BLM administrators' repeatedly expressed their disengagement with the negotiation process concerning Surface Owner Agreements. The perspective of BLM administrators stems from their inability to intervene legally on behalf of either party involved in the negotiation process. However, if an agreement cannot be reached, government regulations allow for the energy developer to "bond on" accessing the privately owned surface to drill.

The practice of bonding on with the BLM allows unfettered access to the mineral estate. Thus, surface owners cannot exercise the property right of exclusion. Within the bundle of sticks that compose property rights, one of the most important is the right of exclusion. And because split-estate property owners cannot wield the stick of exclusion, energy developers have run roughshod over landowners. Simply stated, with the stick comes respect. Additionally, because federal law prohibits split-estate property owners from excluding energy developers, the intervention of government in the negotiation process occurs by default. The implicit nature of government intervention is illustrated by widespread mistreatment of split-estate ranchers and homeowners. The evidence suggests that the practice of bonding on is rare, but landowners are faced with the inevitability of energy development activities taking place. Faced with the knowledge that they cannot prohibit or prevent energy development, split-estate landowners are resigned to making the best deal possible. Government, energy, and ranching interests all described that split-estate property owners as negotiating the best deal possible, and failing that, energy developers could simply bond on with the BLM and access the property. Therefore, evidence points to the conclusion that the BLM is dominated by the energy industry.

Modern-Day Capture of the BLM

The concept of agency's capture is considered by political science as an outdated phenomenon. The case study of split-estate energy development presented here suggests that political science has been too hasty in its dismissal of capture. Agency capture accounts for the centrally important conditions by which the energy industry is able to dominate the policymaking environment of the BLM. While it was clear to previous researchers and commentators that grazing interests had captured the BLM, their capture could last only as long as grazing remained its primary regulatory responsibility (Foss, 1960; Culhane, 1981; Wilkinson, 1992; Cawley, 1993; Clarke & McCool,1996; Klyza, 1996; Starrs, 1998; D. Davis, 1997; Donahue, 1999; Merrill, 2002; Knight, Gilgert & Marston, 2002; Smith & Freemuth, 2007; Nie, 2008). The enactment of the Federal Land Policy and Management Act of

1976 (FLPMA) disrupted the grazing interest iron-triangle of the BLM. Thus, an agency's capture by its most historical definition should not occur again. Thus, the traditional conceptualization of agency capture implies that administrators, no matter the policy domain, act in a manner beneficial to the entity being regulated at the expense of the public good.

The public good becomes expendable as cozy relationships develop between regulated and regulator. The beneficial nature of the cozy relationship between the BLM and the energy industry represents the absence of broader democratic involvement in the decision-making process. An agency's proclivity to utilize democratic principle of civic engagement in the government's decision-making process is of central importance in making the determination of whether or not that agency has been captured or not.

An agency's "modern-day capture" is defined as the administrative *emphasis* of one regulated interest over all other regulated interests. The concept of modern-day capture recognizes that administrative emphasis is beneficial to the regulated entity's activities at the expense of the broader public good. Like the traditional conceptualization of agency capture, modern-day capture also recognizes that a regulated entity's cozy relationship with regulators benefits the regulatory decision-making process. When combined, administrative emphasis and regulatory treatment represent the voices of other interests being drowned out by the most dominant interest. Thus, determination of an agency's modern-day capture suggests that there are degrees by which a regulated entity's benefits come at the expense of all other regulated activities. By definition then, a modern-day capture of an agency is determined by evidence of the agency's being overtly dominated by a regulated entity at the expense of all other regulated activities. Simply stated, modern-day capture of an agency is empirically recognizable.

Industry and elected officials commonly characterize development of domestic energy resources as a public good. As this study indicates, governmental officials and the energy industry articulate that domestic energy development is the "greatest public good" among all other public goods derived from the land. This position suggests that all other use or development of the land is inferior to that of using and developing the mineral estate for its energy resources. This position undermines the economic and social benefits derived by other uses of the surface estate. The view that developing domestic energy resources is the "greatest public good" is a direct reflection of how government officials and the energy industry have come to define what the "greatest public good" means to citizens of the United States.

The view that energy development serves a greater public good is a direct reflection of federal law. In the context of split-estate energy, the legal protection of the mineral estate establishes conditions of unequal footing between the energy industry and those who would make use of the surface estate. The unequal footing, as discussed, is also reflected in the BLM's rules, regulations, procedures, and oversight of split-estate energy development activities. Thus, when the BLM is politically mandated to shift its resources in order to expand domestic energy development, development of those energy resources comes at the expense of the surface-owning public.

Until federal law and regulation are reformed in a manner that restores the legal balance between the use of the mineral estate with the use of the surface estate, energy development interests will retain the upper hand in the policymaking environment of the BLM. Because ranchers did not fully achieve the reforms they were seeking, energy interests continue to control the BLM's land-use subgovernment policymaking environment. This finding implies that the BLM is a land management agency whose policy environment is in a legally defensible state of modern-day capture by the energy industry.

Analysis of federal law and regulation combined with investigative interviews of elites of domestic energy development provides evidence of dominance. Analysis of federal law and regulation shows that the legal dominance of the mineral estate over the surface estate was a significant factor in the ability of the government to shift BLM policy and agency resources favoring energy development. Analysis of evidence tracing the administrative actions taken by President Bush illustrates that his use of executive powers to shift the BLM's energy policy heavily influenced the energy industry's ability to expand its domestic energy activities. Finally, interpretive analysis of interview data demonstrates that the expansion of domestic energy development was at the expense of split-estate landowners. Therefore, because the BLM policy environment currently favors the development of domestic energy resources, a modern-day capture of the BLM by the energy industry has occurred.

Other research efforts in the relationship between government agencies and energy development will likely confirm that modern-day capture remains a useful lens of inquiry to political scientists. One possible avenue for research would be an inquiry into the relationship between the minerals mining service (MMS) and the energy industry. Research of this type would also prove useful as a more general inquiry into the relationship of any public agency having any administrative responsibility for conducting onshore, offshore, or international energy development. Generally speaking, pursuit of an energy research agenda is beneficial to political science and its subfields of study.

The Election of Barak H. Obama

The 2008 election President Barak H. Obama and Vice President Joseph R. Biden evidence indicates that another shift in BLM energy policy is underway. Prior to being sworn in to office, President Obama announced the appointment of Sen. Kenneth L. Salazar (D-CO) as Secretary of the Interior. Immediately following his confirmation, Salazar announced steps to reform BLM energy policy. Secretary Salazar used a Secretarial Order to establish an energy reform team to identify and oversee energy reforms and issued immediate directives to the BLM, announcing to federal administrators that "the BLM will ensure that they, not industry, will determine where, when and how oil and gas leasing will occur" (Dickson, 2010). Following these actions, Salazar declared that the BLM would no longer be the energy industry's "candy store" ("No more," 2010). Secretary Salazar also announced that the administration was taking the necessary steps to "conduct more rigorous reviews of proposed energy leases and permits to drill, increase its consultation with other public agencies, and allow for more public input in future drilling decisions" ("No more," 2010). In its actions and pronouncements concerning reform of BLM energy policy, the Obama administration was making clear that a new policy objective was being strategically pursued by the newly elected president: expanding alternative energy resource development.

Analysis of the evidence presented throughout the case study uncovered collective concern among all groups of interview participants that the new direction in energy policy was troubling. Governmental officials noted that the newly elected administration's pursuit of expanding alternative energy development would have profound impact on their ability to balance multiple use of the land and resources. Their collective unease over expanding alternative energy development is best expressed by Lynn Rust, "The next generation of politically-oriented land use management mandates will take the form of solar arrays, wind farms, geothermal extraction facilities, and mirrored solar towers" (L. Rust, personal communication, May 19, 2009).

During his campaign, Obama announced that once elected, his administration would expand alternative energy development. In making this promise, President Obama was, like his predecessor, announcing his intentions to shift the BLM's energy policy. As noted earlier, a president's unilateral exercise of executive powers within an existing body of federal law and administrative procedures to achieve a political objective impacts the policy environment and, in turn, disrupts the subgovernment. Should President Obama follow the path taken by President Bush to achieve his own political objective, shifting the BLM's energy policy will again disrupt control over the BLM's policymaking environment. Under these hypothetical circumstances, conflict and competition among the triumvirate interests of energy, ranching, and environment are likely because, as Lynn Rust remarks, "If you use up 64 square miles of public lands to develop a solar farm, do you think that ranchers will be able to graze their cattle, or that energy companies will be able to drill on those same public lands? Not likely" (L. Rust, personal communication, May 19, 2009). In turn, preferential treatment of the alternative energy industry could result in renewed competition for control over the BLM's policymaking environment. In this scenario, alternative energy industry's displacement of traditional energy development's dominance would again result in the modern-day capture of the BLM. Tony Herrell argues that, "the potential for alternative energy projects impacting public lands and resources are even greater than the traditional uses of grazing and energy development. If that is what is on the horizon, then this creates a scenario where again, the big guy resource user knocks off the little guy resource user" (T. Herrell, personal communication, May 20, 2009).

Using President Obama's promise as a means to exemplify how conditions for disrupting a policy subgovernment can repeat itself, BLM administrators are expecting that President Obama will act to affect a shift in the BLM's energy policy. Should the Obama administration affect a significant shift in the BLM's energy policies, President Obama will have (1) unilaterally wielded executive power to pursue the policy objective, (2) gained the support of a Democratic-controlled Congress, and (3) established the support of key subcommittees overseeing the energy policies of the BLM. If all of these conditions are met, President Obama will disrupt the BLM's policymaking subgovernment and displace the dominant interest within it. If this hypothetical scenario were to occur, it would be notable because the evidence of the causal chain of events to achieve the policy objective would mean that (1) the degree that presidents can impact change in a policy environment is greater than expected, (2) the rigor and methodological utility of process tracing and interpretative analysis of elite-actor interview data is validated, (3) the findings of this research effort are generalizable and, therefore, are replicable across land management subgovernment policy environments, and (4) the change in interest group domination of a subgovernment policy is cyclical; therefore, the dynamics of a subgovernment can be modeled. It would follow then that if the dynamics of a subgovernment can be modeled during periods of political upheaval, change to subgovernment policy environments is predictable.

Conclusion: Stabilizing a Subgovernment

Legal protection of the mineral estate is disproportionate to the legal protection of the surface estate. This disparity is the result of a body of federal legislation, most notably the Stock-Raising Homestead Act of 1916 (SRHA). Enforcement of the mineral estate's dominance over the surface estate is furthered as federal regulations guiding split-estate energy development. These regulations, such as those within Onshore Order #1, were promulgated in response to the SRHA of 1916 as well as the Mineral Leasing Act of 1920 (MLA). This body of federal laws and regulations guides the BLM's management and oversight of split-estate energy development. It establishes the federal government's prevailing interest in developing the federally owned mineral estate. This body of law and regulation also mandates the BLM protect the government's interest in developing the mineral estate. It is a mandate that reflects the intent of government to serve the public welfare by protecting its ability to provide energy resources to the nation.

The findings from the analysis presented here indicate that the government's legal and regulatory protection and development interests in split-estate energy resources contradict fundamental principles of property ownership and environmental stewardship. These findings establish the conclusion that unless legislative reform of the mineral estate's legal dominance occurs, governmental attempts to balance and protect the interests of the privately owned surface estate in the face of split-estate energy development are categorically impossible to achieve under the current legal environment. The antiquated nature of federal law and regulation controlling the development of split-estate energy resources are at odds with the legal, political, economic, and technological realities of modern-day domestic energy development. The shared understanding expressed by the government, energy, and ranching officials supports this conclusion. Development of non-traditional energy resources such as coalbed methane would not have been possible without a "perfect storm" of legal, political, economic, and technological conditions all coming together within a relatively short period of time. These events established the conditions for political upheaval in the BLM's land-use subgovernment. Simply stated, if the BLM and federal and state governments are to avoid similar upheaval, conflict, and competition, it is necessary to reform federal law. Reforms of the Stock-Raising Homestead Act of 1916 (SRHA), the Mineral Leasing Act of 1920 (MLA), and Onshore Order #1 are required if the government desires to ensure stability in BLM land management policy.

References

Cawley, R. M. (1993). *Federal land western anger: The sagebrush rebellion and environmental politics*. Lawrence, KS: University of Kansas Press.

Clarke J. N. & McCool, D. C. (1996). *Staking out the terrain: Power and performance among natural resource agencies* (2nd ed.). Albany, NY: State University of New York Press.

Culhane, P. J. (1981). *Public lands politics: Interest group influence on the forest service and the bureau of land management*. Baltimore, MD: Johns Hopkins University Press for Resources for the Future.

Davis, D. H. (1997). Energy on federal lands. In C. Davis (Ed.), *Western public lands and environmental politics* (pp.141-168). Boulder, CO: Westview Press.

Dickson, J. (2010, January 8). *Oil and gas leasing reforms arrive: Our efforts have paid off!*. Press release. Retrieved from http://wilderness.org/content/oil-and-gas-leasing-reforms.

Donahue, D. L. (1999). *The western range revisited: Removing livestock from public lands to conserve native biodiversity*. Norman, OK: University of Oklahoma Press.

Foss, P. O. (1960). *Politics and grass: The administration of grazing on the public domain*. Seattle, WA: University of Washington Press.

Herrell, T. (2009). Deputy State Director of Minerals and Lands New Mexico Bureau of Land Management. -Interview Conducted: May 20, 2009; Albuquerque, NM.

Klyza, C. M. (1996). *Who controls public lands?: Mining, forestry, and grazing politics 1870-1990*. Chapel Hill, NC: The University of North Carolina Press.

Knight, R. L., Gilgert W. C., & Marston, E. (Eds.). (2002). *Ranching west of the 100th meridian: Culture, ecology, and economics*. Washington, DC: Island Press.

McCool, D. C. (1989). Subgovernments and the impact of policy fragmentation and accommodation. *Policy Studies Review, 8*(4), 264–287. https://doi.org/10.1111/j.15411338.1988.tb01101.x

McCool, D. C. (1990). Subgovernments as determinants of political viability. *Political Science Quarterly, 105*(2), 269–293.

McCool, D. C. (1995). *Public policy theories, models, and concepts: An anthology*. Englewood Cliffs, NJ: Prentice Hall.

McCool, D. C. (1998). The subsystem family of concepts: A critique and proposal. *Political Research Quarterly, 51*(2), 551–570.

Merrill, K. R. (2002). *Public lands and political meaning: Ranchers, the government, and the property between them*. Berkeley, CA: University of California Press.

Nie, M. (2008). *The governance of western public lands*. Lawrence, KS: University Press of Kansas.

Rust, L. (2009). Deputy State Director of Minerals and Lands Colorado Bureau of Land Management. -Interview Conducted: May 19, 2009; Denver, CO.

Smith, Z. A. & Freemuth, J. C. (Eds.). (2007). *Environmental politics and policy in the west: revised edition*. Boulder, CO: University Press of Colorado.

Starrs, P. F. (1998). *Let the cowboy ride: Cattle ranching in the American west*. Baltimore, MD: The Johns Hopkins University Press.

Unnamed (2009). Former Senior Department of Interior Political Appointee under former President George W. Bush. -Interview Conducted: May 26, 2009; Salt Lake City, UT

Vincent, John (2009). Legal Counsel to the Landowners Association of Wyoming (LAW), Former Mayor of Riverton, WY. -Interview Conducted: March 16, 2009; Riverton, WY

Wilkinson, C. F., (1992). *Crossing the next meridian: Land, water, and the future of the West*. Washington, DC: Island Press.

Index

© Springer Nature Switzerland AG 2019
R. E. Forbis Jr., *Altered Policy Landscapes*,
https://doi.org/10.1007/978-3-030-04774-0